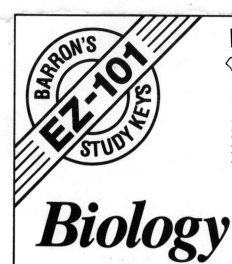

BARRON'S EZ-101 STUDY KEYS

Eli C. Minkoff, Ph.D.,
Professor of Biology
Bates College

Biology

BARRON'S

All inquiries should be addressed to:
Barron's Educational Series, Inc.
250 Wireless Boulevard
Hauppauge, New York 11788

Library of Congress Catalog Card No. 90-27724

International Standard Book No. 0-8120-4569-6

Library of Congress Cataloging-in-Publication Data
Minkoff. Eli C.
   Study keys to biology / Eli C. Minkoff.
       p.   cm.
   Includes index.
   ISBN 0-8120-4569-6
   1. Biology—Study and teaching.   2. Biology.   I. Title.
QH315.M62   1991
574'.076—dc20   90-27724
                                                    CIP

PRINTED IN THE UNITED STATES OF AMERICA

1234   5500   98765432

CONTENTS

Theme 1 BIOLOGY AS A SCIENCE

*B*iology is the study of living systems. Characteristics of these systems include use of energy (metabolism), self-regulating mechanisms (homeostasis), population structure, heredity, and reproduction. Living systems are studied in various ways. Biology is considered scientific because these methods include the testing of falsifiable hypotheses. Great biologists of the past whose work is still influential include Harvey, Linnaeus, Pasteur, Darwin, Mendel, Morgan, Watson, Crick, and others.

INDIVIDUAL KEYS IN THIS THEME
1 What is biology?
2 Scientific methods
3 Landmarks in the history of biology

Key 1   What is biology?

OVERVIEW   *Biology is the science of living systems. These systems exhibit metabolism, irritability, homeostasis, growth, population structure, reproduction, hereditary information, and mutability.*

Living systems All exhibit certain characteristics of life:
- **Metabolism:** Energy-rich materials are broken down; the energy is used or stored, and low-energy wastes are discarded.
- **Selective response (irritability):** All living things can distinguish certain external stimuli from others and can respond selectively.
- **Homeostasis:** Living systems keep themselves going by removing or reducing many harmful conditions and restoring balance.
- **Growth** and **biosynthesis:** Living organisms add material to themselves.
- **Reproduction:** Living organisms make others like themselves.
- **Populations:** Organisms belong to populations of similar organisms.
- **Hereditary information:** All organisms carry hereditary information, derived from parents and other ancestors and capable of mutation.

Mechanism: View of life as just a complex form of physics and chemistry. Most scientists today adopt this strategy in their research.

Vitalism: View that living systems must include "something else," such as a "vital force."

Reductionism: View that attempts to explain wholes in terms of their parts and biology in terms of physics and chemistry.

Compositionism (=holism): View that biology cannot be predicted from physics and chemistry because new phenomena **emerge** when smaller components are put together into complex systems. *Example:* Nothing in chemistry can explain why insects should have six legs or why kangaroos live only in Australia.

Key 2   Scientific methods

OVERVIEW   *Science is a method of inquiry that sets up falsifiable hypotheses and tests them. No statement is scientific unless it is falsifiable. Methods used to test hypotheses may be either experimental or naturalistic or both.*

Science: A method of investigation that proposes testable statements (**hypotheses**) and then subjects the statements to rigorous testing. These statements must be **falsifiable**—capable of being proven false, at least in theory, but not necessarily capable of being proven true.

Scientific method: May be either experimental or naturalistic, but always requires four steps:
1. Facts are gathered.
2. A problem is stated.
3. A **hypothesis** is proposed. (The hypothesis is basically an intelligent guess, but it must be falsifiable.)
4. The hypothesis is tested. (This step requires rigor and objectivity.)

Theory: A hypothesis repeatedly tested and never falsified becomes a **theory**.

Experiment: In an **experiment**, the scientist controls as many variables as possible, allowing only one to vary at a time. If possible, comparison is made between an **experimental group** and a **control group**, which differ only in the one factor being tested. Experimental methods are commonly used in physics, chemistry, physiology, and genetics.

Naturalistic observations: Used for phenomena that cannot be experimentally controlled. Astronomers or paleontologists cannot experiment with distant stars or extinct dinosaurs, and social scientists cannot practically or ethically manipulate society. They are restricted to those experiments that nature performs for them, and must wait patiently for the right conditions to occur. Ecologists, biogeographers, and comparative anatomists often use naturalistic methods.

Key 3 Landmarks in the history of biology

OVERVIEW *Aristotle was the greatest biologist of antiquity; many generations read his works uncritically. Direct observation increased in the Renaissance. Overseas explorations and the invention of the microscope brought many new discoveries. The 19th century brought Pasteur's microbiology, Darwin's evolutionary theory, and Mendel's genetics. Later progress in genetics led to the discovery of the structure of DNA.*

5th cent. B.C. Hippocrates—father of medicine

4th cent. B.C. Aristotle—greatest biologist of antiquity

1350–1650 Renaissance: revival of interest in direct observation, especially in botany, anatomy, and medicine

1600s Descartes—father of experimental physiology

1605 Harvey demonstrates circulation of the blood

1666 Redi disproves spontaneous generation of flies

1600–1800 Age of exploration—discovery of many new organisms

c.1700 Leeuwenhoek invents first practical microscope

mid-1700s Linnaeus—father of taxonomy

1780s Galvani's "animal electricity"—start of nerve physiology

1790s Priestley's work on photosynthesis

1859 Darwin—*Origin of Species*; evolution by natural selection

1860s Pasteur—explains fermentation; discovers many new bacteria

1865 Mendel—father of genetics, (unrecognized until 1900)

1910 Morgan begins work with fruit flies and discovers sex linkage

1910s Bridges and Sturtevant develop genetic mapping technique

1940s Modern synthetic theory of evolution

1953 Watson and Crick describe DNA double helix

1968 Genetic code deciphered

Theme 2 THE CHEMICAL BASIS OF LIFE

*A*ll living systems are built of matter composed of atoms and molecules. Many important characteristics of living systems are determined at the chemical level. Biologically important molecules include lipids, carbohydrates, proteins, and vitamins.

Key 4 Atoms

OVERVIEW *Atoms are the building blocks of all matter. Each atom has a central **nucleus** containing protons and neutrons, surrounded by a cloud of electrons. Atoms can lose or gain electrons and become electrically charged, or they can share electrons.*

Structure of the nucleus:
- Contains **protons** (charge $+1$, 1 mass unit) and **neutrons** (neutral, no charge, 1 mass unit).
- The **atomic number** is the number of protons; the **atomic mass (or "weight")** is the total number of protons and neutrons.
- **Isotopes:** Atoms differing only in the number of neutrons (and thus in atomic mass) are called **isotopes**.

Electron configurations: A cloud of **electrons** (charge -1, mass near 0) surrounds the nucleus. Electrons are arranged in **orbitals** (=**shells**) at different energy levels and distances from the nucleus.
- Shell #1 can take 2 electrons to be full.
- Shells 2 and 3 can take 8 electrons each.
- Shells 4 and 5 can take 18 electrons each.
- A full outer shell means a stable configuration.

Ions and bonds: Electrically charged atoms are called **ions**; they form **ionic bonds**.
- **Metallic elements** tend to *lose* electrons and form positive ions. *Examples:* H^+ hydrogen, Na^+ sodium, Ca^{+2} calcium, K^+ potassium.
- **Nonmetals** tend to *gain* electrons and form negative ions. *Examples:* Cl^- chloride, OH^- hydroxide.

Covalent bonds: Certain atoms can also *share* electrons with other atoms to form **covalent bonds**. *Examples:* H hydrogen (1 bond), O oxygen (2 bonds), N nitrogen (3 bonds), C carbon (4 bonds).

Key 5   Simple molecules

OVERVIEW   *The smallest particle of any compound is called a* **molecule**. *Positive and negative ions attract one another to form ionic compounds. Atoms that share electrons form covalent compounds. Organic compounds are compounds containing covalently bonded carbon.*

Ionic compounds: Form when positive ions **(cations)** and negative ions **(anions)** attract one another electrically.

Covalent bonds: Form when electrons are shared. The element carbon (C) forms covalent bonds in most of its compounds.

Organic compounds: Contain covalently bonded carbon. They are far more numerous than inorganic compounds because carbon atoms can bond covalently into long chains or other shapes.
  • **Hydrocarbons:** Contain hydrogen (H) and carbon (C) only.
  • **Carbon dioxide** (CO_2): Produced by organisms as a product of respiration and used by plants in photosynthesis.

Important functional groups: alcohols ($-OH$), aldehydes ($-CHO$), ketones ($-CO-$), acids ($-COOH$), amines ($-NH_2$), amino acids (both $-NH_2$ and $-COOH$).

KEY MOLECULE: WATER

Water (H_2O): Has a bent shape, with oxygen (O) in the middle.

Water is a **polar covalent** molecule: there is a slight $-$ charge at the oxygen and a slight $+$ charge between the hydrogens.

Water can form weak **hydrogen bonds** with other polar molecules, including other molecules of water. This gives water a surface tension and a high heat capacity.

Water molecules tend to surround dissolved ions.

Water reaches its greatest density at 4°C. Water at 0°C is lighter and floats on top, so ponds tend to freeze on the top only, allowing life to survive in the liquid water underneath. (Most other compounds freeze from the bottom up.)

Key 6 Lipids

OVERVIEW *Lipids are defined by their solubility in solvents like ether. The most important lipids are fats and oils, sources of stored energy made from glycerol and fatty acids.*

KEY BUILDING BLOCKS:

Glycerol:

OH OH OH
| | |
H—CH—CH—CH—H

Isoprene: a branched hydrocarbon, C_5H_8, with two double bonds.

Fatty acids: −COOH at the end of a long hydrocarbon chain.
Saturated fatty acids contain single bonds only;
Unsaturated fatty acids contain some double bonds.

Fats and oils:
- **Monoglycerides and diglycerides** (glycerol + 1 or 2 fatty acids)
- **Triglycerides** or **fats** (glycerol + 3 fatty acids): Those liquid at room temperature are often called **oils**. One water molecule splits out (**dehydration**) for each fatty acid added to glycerol. Fat digestion adds water and splits the molecule (**hydrolysis**). Fats have more calories per gram than most other compounds.

Other lipids:
- **Phospholipids** (like phosphatidyl choline): Occur mostly in membranes.
- **Sphingolipids** (like sphingomyelin): Have sphingosine instead of glycerol.
- **Glycolipids** (like cerebrosides and gangliosides in the brain): Have sugar instead of glycerol.
- **Waxes:** Are long-chain alcohols found in plants.
- **Terpenes:** Are odorous hydrocarbons in plants, built from isoprene units.
- **Carotenoids** (carotenes and xanthophylls): Are plant pigments composed of isoprene units.
- **Steroids:** Include **cholesterol** and its derivatives (Key 51).

Key 7   Carbohydrates

OVERVIEW *Carbohydrates, including simple sugars (monosaccharides), double sugars (disaccharides), and polysaccharides like starch, are the major source of energy for most organisms. Proteins and other biological compounds can sometimes have important carbohydrates attached to their surfaces.*

Monosaccharides (simple sugars): Many can change from a straight chain to a ring, and back again. Some are aldehydes ($-CHO$); others are ketones ($-CO-$). All have multiple $-OH$ groups.
  Examples of 5-carbon sugars (pentoses): ribose, deoxyribose.
  Examples of 6-carbon sugars (hexoses): glucose, fructose, galactose.

Disaccharides (double sugars): Can be formed by dehydration (splitting out water) between two monosaccharide molecules. *Examples:* sucrose (table sugar, =glucose + fructose); lactose (milk sugar, = glucose + galactose); maltose (malt sugar, = glucose + glucose).

Polysaccharides: Can be formed by dehydration (splitting out water), joining many monosaccharides into long chains which may sometimes branch. *Examples:* glycogen (used for storage in animals); starch (used for storage in plants); cellulose (found in plant cell walls). Some polysaccharides can be bonded to protein, forming protein-polysaccharide complexes.

Key 8 Proteins in general

OVERVIEW *Proteins are molecules that serve many important biological functions. **Structural proteins** are important in building cells. Other proteins function as enzymes (Key 9), hormones (Key 50), antibodies (Key 42), or oxygen-carrying molecules.*

Amino acids: Proteins are chains of **amino acids**, each containing carbon, hydrogen, oxygen, and nitrogen, and sometimes sulfur. Each amino acid contains an amino group ($-NH_2$) and an organic acid group ($-COOH$), both bound to the same central carbon.
- Each amino acid has a different group (commonly symbolized as $-R$) attached to the central carbon.
- Twenty amino acids commonly occur in proteins.
- About half of the amino acids are **essential** for human nutrition because we cannot synthesize them ourselves; we need them in our diet. Most animal proteins, but few plant proteins, have all the essential amino acids; they are called nutritionally **complete**.

Primary protein structure: A chain-like sequence of amino acids joined by **peptide bonds**. When these bonds form, water is split out (dehydration). Short amino acid chains are called **oligopeptides**; long chains are called **polypeptides**.

Secondary protein structure: Shapes held together by weak **hydrogen bonds**. **Alpha helix** (most common): 3 amino acids to each 360° turn. **Beta pleat:** sheet with zig-zag folds.

Tertiary protein structure: Overall 3-dimensional shape, as determined by **disulfide bridges** linking polypeptide chains together or by electrical forces between positive and negative charges. Many proteins are folded into compact, **globular** shapes.

Quaternary protein structure (in some proteins only): Cannot function (as enzymes or oxygen carriers) unless two or more tertiary structures combine.

Conjugated proteins: Contain a sugar or lipid component.

Key 9 Proteins as enzymes

OVERVIEW *Enzymes are organic molecules that speed up chemical reactions without being used up or altered. All enzymes are proteins, but some have nonprotein portions as well. Enzymes can be used to regulate biochemical pathways. Enzyme molecules are recycled, and so are needed in very small amounts only.*

Catalyst: Something that speeds up a chemical reaction, but is not itself used up or altered as a result.

Enzyme: An organic catalyst, made largely or entirely of protein.

Substrate: A chemical substance upon which an enzyme acts. Many enzymes are named after their substrate plus *-ase*. Example: the enzyme *lactase* digests lactose, a sugar.

Structure: Every enzyme has an **active site** that must bind to the substrate; small structural changes here have large effects and may inactivate the enzyme. The rest of the enzyme holds the active site in the proper position; small changes here are usually less critical.

Isoenzymes: Most enzymes exist in several varieties, called **isoenzymes** or **isozymes**, which function alike but can be distinguished using **electrophoresis**, a technique that separates proteins based on their speed of migration through an electric field.

Function: Enzymes speed up the *rate* of the reaction they catalyze. The enzyme binds to the substrate, forming an **enzyme-substrate complex**. Reaction products are then released, and the enzyme molecule is recycled to react with another substrate molecule. Most enzymes function best at a particular pH or temperature optimum.

Specificity: Most enzymes match their substrates with a **lock-and-key fit**, and are thus highly **specific** to a particular substrate. Some substrates may change the shape of their enzymes to make them fit (the **induced fit hypothesis**).

Allostery: Enzyme activity often depends on a very slight freedom of controlled movement called **allostery** or **allosteric control**.

Inhibitors: They can inactivate enzymes by attaching to the active site, by changing the shape of the enzyme molecule, or by interfering with allostery. All enzymes can be destroyed by **denaturing** their proteins (as by boiling).

Key 10 Vitamins and cofactors

OVERVIEW *Certain enzymes need very small amounts of nonprotein components (cofactors) such as metallic ions or organic molecules called coenzymes. Vitamins are organic nutrients required in very small quantities; many vitamins are known to function as coenzymes.*

Cofactors: Many enzymes (and also hemoglobins and chlorophyll) cannot function properly unless a particular metallic ion (like Fe^{+3} or Mg^{+2}) is present as a **cofactor**. *Examples:* iron in the bacterial enzyme *catalase* or in hemoglobin; magnesium in chlorophyll.

Coenzymes: Some enzymes need organic cofactors (**coenzymes**). *Examples:* Malate dehydrogenase (in the Krebs cycle) requires the coenzyme NAD^+; hemoglobin requires a ring-like "heme" group.

Vitamins: Organic nutrients needed in our diets in very small quantities. Many vitamins function as coenzymes.

Fat-soluble vitamins: Can accumulate in the body's fat reserves; chronic overdoses can thus build up to toxic levels.
- **Vitamin A.** Derived from carotene; forms **rhodopsin**, a light-collecting pigment in the retina.
- **Vitamin D.** Prevents rickets; regulates calcium and phosphorus metabolism. Final step in synthesis uses sunlight.
- **Vitamin E.** An anti-oxidant which protects membranes.
- **Vitamin K.** Binds calcium ions; needed for blood coagulation.

Water-soluble vitamins: Circulate in body fluids. Overdoses are unlikely because excess quantities are excreted in the urine.
- **Ascorbic acid (Vitamin C).** An anti-oxidant; promotes healthy mucous membranes and collagen; prevents scurvy.
- **Thiamine (Vitamin B_1).** Needed in carbohydrate metabolism.
- **Riboflavin (Vitamin B_2).** Part of the coenzymes FMN and FAD.
- **Niacin (Vitamin B_3).** Helps make NAD and NADP coenzymes.
- **Pantothenic acid.** Needed to make coenzyme A (Key 19).
- **Other B-group vitamins:** folacin (folic acid + folate), pyridoxine (B_6), cyanocobalamin (B_{12}), biotin.

Theme 3 CELLS AND TISSUES

*L*iving organisms are all made of cells. Each cell contains several types of organelles and each is surrounded by a membrane. Most cells have a central nucleus containing the chromosomes, which carry hereditary information. Normal cell division (mitosis) ensures that each new cell carries all the hereditary information of its parent cell. A special form of cell division called meiosis gives rise to sex cells (gametes) that then combine during fertilization. Most multicellular organisms are constructed of groups of similar cells called tissues.

Key 11 Cells: Basic structure

OVERVIEW *Bacteria and cyanobacteria have simple (**procaryotic**) cells without a well-defined nucleus. All other organisms have **eucaryotic** cells, divided into a membrane-limited **nucleus** and a surrounding **cytoplasm**.*

Procaryotic cells (found in bacteria and blue-green cyanobacteria): Small ($1-10$ μm); no well-defined nucleus; no internal membranes or membranous organelles (endoplasmic reticulum, Golgi apparatus, vacuoles, lysosomes, or vesicles); no mitochondria or plastids; no centrioles, microtubules, spindle, or true mitosis or meiosis; only one chromosome, containing nucleic acids only; no multicellular organization.

Eucaryotic cells (found in all other organisms): Large ($10-100$ μm); contains well-defined nucleus, surrounded by a nuclear membrane; many internal membranes, arranged into membranous organelles (endoplasmic reticulum, Golgi apparatus, vacuoles, lysosomes, vesicles); mitochondria always present; plastids occur in all plants; centrioles (with "9 + 2" structure) and microtubules; true mitosis (using spindle fibers) and meiosis; multiple chromosomes, containing proteins (histones) in addition to DNA; frequent multicellular organization. All eucaryotic cells divided into:
- **Nucleus:** contains chromosomes; responsible for heredity
- **Cytoplasm:** surrounds nucleus; responsible for cell metabolism

Major types of eucaryotic cells:
- **Animal cells:** No cell walls; many small vacuoles; nucleus in center; centrioles present; no plastids or chlorophyll.
- **Plant cells:** Surrounded by cell wall containing cellulose; one large vacuole; nucleus off to one side; chloroplasts (containing chlorophyll) and other plastids; often no centrioles.

Cell theory (first suggested by Schleiden and Schwann around 1830): All organisms are composed of cells. All substances produced by organisms are made by cells. All cells arise from pre-existing cells.

Key 12   Membrane organelles

OVERVIEW   *The cytoplasm of eucaryotic cells contains various organelles made largely of membranes. Such membranes include the plasma membrane (cell membrane) lining the outside of the cell, as well as a transport system which includes the endoplasmic reticulum, Golgi apparatus, vacuoles, lysosomes, and vesicles. In all cases, the membrane is composed of a lipid bilayer with various proteins attached. Transport across these membranes may be either passive transport along a concentration gradient or active transport, a process which requires energy. The mass transport of large quantities of material is always an active, energy-using process. Membrane structure can be summarized as "protein icebergs in a sea of grease."*

Membrane structure: All cell and organelle membranes have the same basic structure **(unit membrane hypothesis)**. A major component of every membrane is a **lipid bilayer**; some of the lipids here are **phospholipids**. The long non-water-soluble (hydrophobic) hydrocarbon chains of the two layers interlace with (or dissolve in) one another, while the polar (hydrophilic, water-soluble) "heads" of the molecules form the inner and outer membrane surfaces. Most membranes also contain **proteins**, either as **integral proteins** (part of the membrane) or **surface proteins**. Some proteins may snake through the membrane from one surface to another. Some membranes have **glycoproteins**, containing carbohydrate components that stick out from the surface. Surface proteins and glycoproteins often aid in cell recognition.

Organelles made primarily of membranes:
- **Cell membrane (plasma membrane):** Outer lining of each cell; protects the cell and controls what can enter and leave.
- **Endoplasmic reticulum (ER):** Series of folded membranes important in the transport of materials within the cell. May be either **rough** if ribosomes (Key 13) are attached, or **smooth** if they are not. Cavities within the folds of the ER are called **cisternae**.
- **Golgi apparatus:** Made of stacked and somewhat flattened vesicles important in the packaging of proteins into vacuoles.
- **Nuclear envelope:** Two-layered membrane around nucleus, penetrated by **nuclear pores**.

- **Vacuoles:** Spherical droplets surrounded by a membrane, often formed by budding off the margins of the Golgi apparatus. Contents may include fat, protein, ingested food, or a secretion product.
- **Lysosomes:** Organelles containing protein-digesting enzymes surrounded by a membrane. Food vacuoles generally fuse with lysosomes as part of their digestion. Rupture of the lysosomes causes **autolysis**, a process in which the cell digests itself.
- **Vesicles:** Spherical or teardrop-shaped sacs formed by pinocytosis or phagocytosis.

Passive transport (diffusion): Transport along a **concentration gradient** (from a region of higher concentration to one of lower concentration), using no energy.

Active transport: Transport *against* the concentration gradient, requiring energy.

Bulk transport (always requiring energy):
- **Exocytosis:** Secretion or release of material from cells in which vacuole fuses to plasma membrane, releasing contents.
- **Endocytosis:** The reverse process, in which vacuoles are formed:
  1. **Phagocytosis:** Formation of large vacuoles by folds of the cell membrane engulfing material from outside the cell.
  2. **Pinocytosis:** Formation of smaller vacuoles at the bottoms of funnel-shaped depressions in the cell membrane.

Key 13 Other organelles

OVERVIEW *Cytoplasmic organelles not already described include mitochondria, ribosomes, and a cytoskeleton containing microfilaments and microtubules, of which some are organized into cilia and centrioles. Plant cells also have several kinds of plastids, including the chloroplasts responsible for photosynthesis.*

Mitochondria: Produce energy (Keys 19–20). Each has an **outer membrane** (not folded) and an **inner membrane** whose folds are called **cristae**. The interior is filled with a **matrix** containing both enzymes and **mitochondrial DNA**.

Ribosomes: Small units essential in protein synthesis (Key 29). Each has a small (30S) subunit and a larger (50S) subunit. May occur alone, but in eucaryotes often attach to the endoplasmic reticulum (Key 12).

Cytoskeleton and tubular organelles (in eucaryotic cells only):
- **Microfilaments:** Protein fibers (**actin, myosin**, etc.) used in phagocytosis (Key 12), amoeboid motility, and muscle contraction (Key 45).
- **Microtubules:** Hollow structures made of protein, important in cellular motility, and used in mitosis (Key 14) as spindle fibers.
- **Centrioles, cilia, and eucaryotic flagellae:** Made of microtubules arranged as nine pairs or triplets in a circle, often surrounding two single tubules.

Plastids (in plant cells only):
- **Leucoplasts:** Colorless plastids used in food storage.
- **Chloroplasts:** Green plastids with chlorophyll; carry out photosynthesis (Key 17). Contain **grana** (stacks of disk-shaped **thylakoids**) and **stroma** (spaces between the grana).
- **Chromoplasts:** Plastids with pigments other than chlorophyll.

Key 14   Nucleus and mitosis

OVERVIEW   *The nucleus of each eucaryotic cell contains chromosomes that control heredity. These chromosomes replicate in the **cell cycle**, which includes mitosis, cytokinesis, and interphase.*

Nuclear envelope: Two-layered membrane around nucleus.

Nucleolus: Dense-staining, RNA-rich region within the nucleus, responsible for the production of ribosomes (Key 13).

Chromatin: Long, thin, threads of DNA, hard to see except when they appear as **chromosomes** during cell division.

Cell cycle: Consists of nuclear division (mitosis), cytoplasmic division (cytokinesis), and interphase; results in two identical cells.

Mitosis: Division of the nucleus in **somatic cells** (all cells except sex cells). It *preserves chromosome number without change*.
- **Prophase:** Centrioles divide and separate; spindle forms between them, then enlarges. Nuclear membrane and nucleolus fade from view. Chromosomes thicken and coil more tightly, becoming visible as two **chromatids**, attached at a central **centromere**.
- **Metaphase:** Chromosomes line up on the equator of the cell to form a **metaphase plate**.
- **Anaphase:** Centromeres of each chromosome divide in half, briefly doubling the chromosome number (since we count centromeres). **Sister chromatids** (those derived from the same chromosome) now separate and are pulled toward opposite poles.
- **Telophase:** Chromosomes reach opposite poles and grow too thin to see. Finally, nuclear membrane and nucleolus reappear.

Cytokinesis (division of the cytoplasm, much simpler than mitosis).
- In animal cells (and other cells without rigid cell walls), cytoplasm divides by constricting in the middle, forming a dumbbell shape.
- In cells with rigid cell walls (like plant cells), a series of bubble-like vesicles fuse to form a **cell plate** along the cell's equator.

Interphase: The longest part of each cell cycle, during which nearly all growth and metabolism occurs. The three phases of interphase are:
- G_1 (first gap phase): time of most rapid growth; many new organelles synthesized.
- S (synthesis phase): DNA synthesis occurs; growth continues.
- G_2 (second gap phase): growth continues, chromosomes become double-stranded; cell prepares to divide.

Key 15 Meiosis and fertilization

OVERVIEW *Meiosis is a special kind of cell division that produces haploid sex cells (**gametes**) after two nuclear divisions. The diploid number is restored in fertilization when male and female gametes unite.*

Chromosome numbers: A **haploid (N)** set of chromosomes contains one chromosome of each type. Two haploid sets make a **diploid** number **(2N)**. In **diploid** cells, all chromosomes exist in **homologous pairs**. Familiar animals and plants have diploid cells except for gametes.

Meiosis: A special type of cell division in which diploid cells divide twice to produce haploid sex cells **(gametes)**. Stages of meiosis include:

- **Prophase I:** Chromosomes first become visible to light microscopy **(leptotene)**; homologous chromosomes pair up **(zygotene)**; chromosomes thicken **(pachytene)** and crossing-over may occur; individual chromatids become visible to light microscopy **(diplotene)**, with chromosomes arranged in **tetrads** (= 2 homologous chromosomes or 4 chromatids); chromosomes of each homologous pair begin to separate a bit **(diakinesis)**, with an X-shaped **chiasma** marking the location of each crossover.
- **Prometaphase:** Spindle forms and enlarges; centrioles move to opposite poles; nuclear membrane dissolves.
- **Metaphase I:** Paired chromosomes line up on metaphase plate.
- **Anaphase I:** Homologous chromosomes separate and move toward poles.
- **Telophase I:** Homologous chromosomes reach opposite poles but remain condensed (short and thick).
- **Interkinesis** (may not occur): Cytoplasm divides in half.
- **Prophase II:** New spindles form, perpendicular to that of division 1.
- **Metaphase II:** Chromosomes line up on new metaphase plate.
- **Anaphase II:** Centromeres divide; sister chromatids separate as in mitosis, forming separate chromosomes.
- **Telophase II:** New chromosomes reach opposite poles and begin to lengthen and thin out; nuclear membrane reappears.
- **Cytokinesis:** Cytoplasm divides, resulting in four haploid cells.

Spermatogenesis (meiosis in males): Cytoplasmic divisions are equal, resulting in four equal cells. **Primary spermatocytes** (diploid) undergo first division of meiosis, producing haploid **secondary spermatocytes**. Secondary spermatocytes then undergo second division of meiosis; four haploid **spermatids** result. Each spermatid undergoes maturation (growth of tail, loss of most cytoplasm) to become a mature **sperm cell (spermatozoon)**.

Oogenesis (meiosis in females): Cytoplasmic divisions are very unequal, resulting in one ovum only and several smaller cells (polar bodies). In first division, **primary oocyte** (diploid) divides into **secondary oocyte** (haploid) and **first polar body** (haploid). Secondary oocyte then divides into mature **egg cell (ovum)** and **second polar body** (both haploid).

Fertilization: Small haploid sperm penetrates membrane surrounding large egg. Sperm tail stays outside; only sperm head (with nucleus) enters.

- **Primary (fast) block:** Wave of negative electric polarity spreads across surface of egg from point of sperm entry, reducing (but not eliminating) probability of a second sperm entering **(polyspermy)**.
- **Secondary (slow) block:** Vesicles pop open, separating membrane from egg, reducing probability of polyspermy to nearly zero.
- Haploid **sperm pronucleus** approaches haploid **egg pronucleus**; they fuse to produce diploid nucleus of fertilized egg **(zygote)**, which immediately begins mitosis (first cleavage).

Key 16 Tissues

OVERVIEW *Tissues are groups of similar cells closely related in function and location. All animals and plants have cells organized into tissues. In animals, the four major types of tissues are epithelium, connective tissue, muscle tissue, and nervous tissue.*

Epithelium: Tissues that originate in broad, flat surfaces; functions include protection, absorption, and secretion.
- **Simple** (one cell layer thick); may be **squamous** (cells flat, e.g., lining of capillaries), **cuboidal** (cells equal in height and width), or **columnar** (cells tall and skinny, e.g., inner lining of intestine).
- **Stratified** (many cell layers thick); may be **squamous** (surface cells flat, e.g., epidermis of skin), **cuboidal** (surface cells equal in height and width), or **columnar** (surface cells tall and skinny).
- **Glandular** (highly modified for secretion); may be **exocrine** (secretions exit by ducts to targets nearby) or **endocrine** (secretions carried by blood stream, targets may be far away, see Key 50).

Connective tissues: Tissues with much extracellular material (**matrix**).
- With liquid matrix (**plasma**), capable of solidifying in a **clot**.
    1. **Blood**. Contains **plasma** (fluid containing water, ions, proteins, etc.); **erythrocytes** (red blood cells containing hemoglobin); **leucocytes** (white blood cells of many types: lymphocytes, monocytes, neutrophils, basophils, eosinophils); and **platelets** (cytoplasmic fragments that aid in clotting).
    2. **Lymph**. Contains leucocytes and plasma only.
- With solid matrix, often with fibers of **collagen** and related proteins.
    1. **Loose (areolar) connective tissue:** Simplest, with few fibers.
    2. **Regular connective tissue:** Fibers all arranged in same direction (as in ligaments and tendons).
    3. **Irregular connective tissue:** Fibers arranged in all directions (as in dermis of skin).
    4. **Adipose (fat) tissue:** Cells filled with fat deposits.
    5. **Hemopoietic (blood-forming) tissue:** Contains precursors to many connective tissue cells (e.g., bone marrow).
    6. **Cartilage:** Contains a shock-resistant complex of protein and sugar-like (polysaccharide) molecules.
    7. **Bone:** Strong tissue containing calcium salts.

Muscle tissues: Tissues specialized for contraction.
- **Smooth muscle:** Involuntary cells with tapering ends but no cross-banding; smooth rhythmic contractions; nuclei located centrally; occurs in digestive organs, reproductive organs, etc.
- **Cardiac muscle:** Involuntary fibers with **cross-striations**; cylindrical in shape, but branching and coming together repeatedly; nuclei located centrally; cell boundaries marked by **intercalated disks**; rhythmic contractions; occurs in heart only.
- **Skeletal muscle:** Voluntary cells with cylindrical shape; **cross-striations** caused by alignment of actin and myosin fibers; many nuclei per fiber, no cell boundaries (each fiber is thus called a **syncytium**); rapid, forceful contractions, but fatigues easily; occurs in muscles; always attaches to connective tissues.

Nervous tissues: Contain specialized nerve cells (neurons).
- **Neurons:** Respond to stimuli by conducting impulses (Key 46).
- **Neuroglia** (several types): Hold nervous tissue together.

Plant tissue types (very different from animals):
- **Epidermis:** Arranged in flat surfaces; usually has a protective function.
- **Meristem:** Embryonic tissues which continue growing throughout life.
- **Supporting tissues** (many types): Hold the plant up and give it shape.
- **Vascular (conducting) tissues:** Transport liquids and help in support.
  1. **Xylem (wood):** Generally carries fluids upwards.
  2. **Phloem:** Carries photosynthetic products downwards.

Theme 4 BIOENERGETICS

*B*iological systems cannot keep going unless energy keeps flowing through them. The energy in all biological systems originates as solar energy and is converted into chemical energy by photosynthesis. Cells use this energy by breaking down substances like glucose into pyruvate. Pyruvate is then broken down further by a repeating series of reactions called the Krebs cycle. The largest amount of energy from these processes is released with the help of an electron transport chain.

Key 17 Photosynthesis

OVERVIEW *Photosynthesis is a process by which plants (and certain bacteria) make sugars with the aid of sunlight. Reactions that require light (**light reactions**) include electron capture by chlorophyll or other pigments and the use of that electron to split a molecule of water and release oxygen. Reactions that do not require light (**dark reactions**) include the fixing of CO_2 and formation of RUDP and later glucose.*

Anabolism: The building up of larger, energy-rich molecules from smaller ones. All anabolic processes require energy.

Photosynthesis: Anabolic process by which plants, some bacteria, and cyanobacteria use sunlight to make glucose and other sugars. Photosynthesis requires a light-capturing pigment, usually **chlorophyll**. In true plants, photosynthesis only takes place in the **chloroplasts**. The overall reaction can be summarized as:

$$6\,CO_2 + 12\,H_2O + energy \longrightarrow C_6H_{12}O_6 \text{ (glucose)} + 6\,O_2 + 6\,H_2O$$

All of the oxygen on the right comes from the water on the left.

Light reactions (light-sensitive, requiring light of certain wavelengths):

$$H_2O + NADP^+ + ADP + P_i + energy \longrightarrow O_2 + ATP + NADPH + H^+$$

The light energy needed for this reaction must be trapped by **pigments**, especially **chlorophyll a** and **chlorophyll b**; oxygen is released and ATP is formed.

Pigments: Chlorophyll a has a light-absorption maximum in the 400–460 nm range (blue) and another between 650 and 700 nm (red). Chlorophyll looks green because it absorbs these other wavelengths. **Accessory pigments** (including carotenes, xanthophylls, and phycobilins) capture light energy at other wavelengths.

Hill reaction: Releases electrons to the photosystems:

$$H_2O + NADP^+ + energy \longrightarrow O_2 + NADPH + H^+$$

Photosystems I and II: Photosynthetic pigments occur in two complexes called **photosystem I** and **photosystem II**. Photosystem II must be excited by a light **quantum**, boosting the energy of an electron from the hydrogen of H_2O. The electron passes from one pigment to another and then to photosystem I, which must be excited by another light quantum in order for the process to continue. Photosystem I transfers the electron to a second series of pigments and ultimately to $NADP^+$, which then enters the dark reactions.

Dark reactions (light-insensitive, do not require light): ATP and NADPH from the light reactions are used in the dark reactions to incorporate CO_2 into plant tissues and produce sugars.

KEY REACTION SERIES: THE CALVIN CYCLE

Most plants use the **Calvin cycle** to fix CO_2 into a 3-carbon compound; they are therefore called **C_3** plants. Steps in this process are:

1. CO_2 + **RUDP** (ribulose 1,5-diphosphate) $\longrightarrow$ 2 molecules 3-phosphoglycerate.
2. 3-phosphoglycerate + ATP then NADPH + H^+ $\longrightarrow$ **G3P** (glyceraldehyde 3-phosphate) + $NADP^+$ + ADP + P_i (inorganic phosphate).
3. Some G3P $\longrightarrow$ **DHAP** (dihydroxyacetone phosphate).
4. Some G3P + DHAP $\longrightarrow$ fructose diphosphate $\longrightarrow$ fructose-6-phosphate + P_i
5. Some fructose-6-phosphate $\longrightarrow$ glucose-6-phosphate $\longrightarrow$ glucose + P_i; or else glucose-6-phosphate $\longrightarrow$ starch + P_i.
6. Rest of fructose-6-phosphate + G3P $\longrightarrow$ xylulose-5-phosphate + erythrose-4-phosphate.
7. Erythrose-4-phosphate + DHAP $\longrightarrow$ sedoheptulose diphosphate $\longrightarrow$ sedoheptulose 7-phosphate + P_i
8. Sedoheptulose 7-phosphate + G3P $\longrightarrow$ xylulose-5-phosphate + ribose-5-phosphate.
9. Ribose 5-phosphate $\longrightarrow$ ribulose 5-phosphate.
10. Xylulose-5-phosphate (from reactions 6 & 8) $\longrightarrow$ ribulose 5-phosphate.
11. Ribulose-5-phosphate + ATP $\longrightarrow$ RUDP + ADP.

Overall net reaction for the Calvin cycle:

$$6\ CO_2 + 18\ ATP + 12\ NADPH + 12\ H^+ + 12\ H_2O \longrightarrow C_6H_{12}O_6(\text{glucose}) + 18\ P_i + 18\ ADP + 12\ NADP^+$$

C_4 plants: Certain plants like corn and sugarcane are called **C_4** plants because **CO_2** is initially fixed into a 4-carbon compound. **C_4** plants are mostly tropical. They can make glucose faster and grow faster than **C_3** plants and use less water, but they use more ATP to make each molecule of glucose.

Key 18   Glucose metabolism

OVERVIEW   *Sugars are converted to pyruvate by* **glycol-ysis**. *Each glucose produces two pyruvate molecules, using two ATPs but synthesizing four (net gain of two). Further reactions depend on whether or not O_2 is present.*

Catabolism: The breakdown of energy-rich molecules such as glucose, a process that releases energy.

ATP: Most biological processes that require energy use **Adenosine tri-phosphate (ATP)**. Energy-producing reactions synthesize ATP from adenosine diphosphate **(ADP)** and inorganic phosphate **(P_i)**. Most ATP comes from the catabolism of glucose, including glycolysis, the Krebs cycle (Key 19), and electron transport (Key 20).

Glycolysis: The breakdown of sugars to pyruvate.

KEY REACTION SERIES: GLYCOLYSIS

1. Glucose + ATP $\longrightarrow$ ADP + glucose 6-phosphate
2. Glucose 6-phosphate $\longrightarrow$ fructose 6-phosphate
3. Fructose 6-phosphate + ATP $\longrightarrow$ ADP + fructose 1,6-diphosphate
4. The six-carbon compound is now split in half: Fructose 1,6-diphos-phate $\longrightarrow$ glyceraldehyde 3-phosphate **(G3P)** + dihydroxyace-tone phosphate **(DHAP)**
5. DHAP $\longrightarrow$ G3P
6. 2 G3P + 2 NAD^+ + 2 P_i $\longrightarrow$ 2 1,3-diphosphoglycerate + 2 NADH + 2 H^+.
7. 2 1,3-diphosphoglycerate + 2 ADP $\longrightarrow$ 2 3-phosphoglycer-ate + 2 ATP
8. 2 3-phosphoglycerate $\longrightarrow$ 2 2-phosphoglycerate
9. 2 2-phosphoglycerate $\longrightarrow$ 2 PhosphoEnolPyruvate (PEP)
10. 2 PEP + 2 ADP $\longrightarrow$ 2 **pyruvate** + 2 ATP

Aerobic metabolism: If oxygen is present, pyruvate forms **acetyl co-enzyme A**, which then enters the Krebs cycle (Key 19).

Anaerobic metabolism: Pyruvate undergoes **fermentation**.
- In microorganisms, ethyl alcohol is formed:
  pyruvate $\longrightarrow CO_2$ + acetaldehyde; acetaldehyde + NADH + $H^+ \longrightarrow$ ethyl alcohol + NAD^+
- In muscles: pyruvate + NADH + $H^+ \longrightarrow$ lactic acid (or lactate) + NAD^+

Key 19 Krebs cycle

OVERVIEW *In the breakdown of glucose and other sugars, most of the ATP is formed in a cyclical series of reactions called the* **citric acid cycle** *or* **Krebs cycle**. *CO_2 is formed in this cycle. Most of the ATP is made indirectly via $NADH^+$ and the electron transport chain (Key 20).*

Production of acetyl coenzyme A (acetyl-CoA):

$$Pyruvate + NAD^+ + Coenzyme\ A \longrightarrow Acetyl\text{-}CoA + NADH + H^+ + CO_2$$

Acetyl-CoA also results from the catabolism of fats and proteins.

KEY REACTION SERIES

Krebs cycle, also called **citric acid** or **tricarboxylic acid (TCA) cycle**

1. Acetyl-CoA + oxaloacetate $\longrightarrow$ citrate
2. Citrate $\longrightarrow$ *cis*-acotinate + H_2O
3. *Cis*-acotinate + $H_2O \longrightarrow$ isocitrate (an isomer of citrate)
4. Isocitrate + $NAD^+ \longrightarrow$ oxalosuccinate + $NADH + H^+$; oxalosuccinate (unstable) $\longrightarrow$ α-ketoglutarate + CO_2
5. α-ketoglutarate + coenzyme A + $NAD^+ \longrightarrow$
   succinyl-Co-A + $NADH + H^+ + CO_2$
6. Succinyl-Co-A + GDP + $P_i \longrightarrow$ succinate + GTP + coenzyme A
   NOTE: GTP (similar to ATP) transfers the phosphate to ATP:
   $$GTP + ADP \longrightarrow GDP + ATP$$
7. Succinate + $FAD^+ \longrightarrow$ fumarate + FADH + H^+
8. Fumarate + $H_2O \longrightarrow$ malate
9. Malate + $NAD^+ \longrightarrow$ oxaloacetate + $NADH + H^+$
   The oxaloacetate then recycles back to reaction 1.

- These reactions are sometimes written in terms of *oxaloacetic acid* or *citric acid*, but in living cells these compounds are usually ionized as citrate, oxaloacetate, etc.
- Many of these reactions are reversible, and are also used in amino acid synthesis.
- Krebs cycle reactions usually take place inside mitochondria, with some of the enzymes bound to the cristae of the inner membrane while others float freely in the fluid matrix.

Key 20   Electron transport system

OVERVIEW   *Throughout the Krebs cycle and glycolysis, hydrogen atoms and their electrons are passed from one energy level to the next by a series of electron acceptors, including NAD and cytochromes. Atmospheric oxygen, the final electron acceptor, combines with hydrogen to make H_2O.*

Cytochromes: Electron acceptors (proteins + porphyrin rings).

Electron transport system (cytochrome system): During catabolism, hydrogen atoms and their electrons are removed by electron acceptors like **Nicotinamide Adenine Dinucleotide (NAD), NAD Phosphate (NADP)**, and **Flavin Adenine Dinucleotide (FAD)**. Electrons then pass down the energy levels in the cytochrome chain:

At site I: NAD or NADP
                  Flavin MonoNucleotide (**FMN**)
                  Iron-sulfur proteins (**ferredoxins**)
                  Coenzyme Q (Ubiquinone)
At site II: Cytochrome *b*
                  Cytochrome *c*
At site III: Cytochrome *a*
                  Cytochrome a_3
                  Oxygen atoms from atmospheric O_2

- In each reaction, one compound is oxidized (loses an electron) and another is reduced (gaining the electron). Example:
Cytochrome *b* (reduced form) $\longrightarrow$ Cytochrome *b* (oxidized form); electron is carried to cytochrome *c* (which becomes reduced).
- The reactions occur at three distinct sites on the inner mitochondrial membrane; one molecule of ATP is formed at each site.
- When glucose breaks down, most of the ATP comes from the electron transport system, but only if O_2 is present.

Chemiosmosis: The reactions at each site pump protons (H^+) into the space between the mitochondrial membranes. Protons flowing back across the inner membrane drive ATP synthesis.

Oxidative phosphorylation: The process for making ATP from ADP. Its rate is usually controlled (limited) by the amount of ADP coming from ATP breakdown, thus by the rate of energy use.

Theme 5 GENETICS AND DEVELOPMENT

*G*enetics is the study of hereditary information. Hereditary material is organized into genes. Transmission of hereditary information can be studied by making various crosses. Through these crosses, geneticists have been able to map where genes are located along the chromosomes, including those on the special pair of chromosomes that determine sex in many species. Some genes cause inherited diseases, and others are expressed in very complex ways. Most genes made of DNA are expressed by copying a DNA sequence into a sequence of messenger RNA and then using that messenger RNA to synthesize a protein. New techniques using recombinant DNA allow geneticists to probe more deeply into gene structure.

Important developmental changes include cleavage, blastula formation, gastrulation, and neurulation. Many of these changes are controlled by chemical substances called organizers. Organizers work by turning protein synthesis on or off, thus controlling gene expression.

INDIVIDUAL KEYS IN THIS THEME

Key 21   Genetics: Simple crosses with one gene

OVERVIEW   *Gregor Mendel discovered that genes are inherited as discrete particles that do not blend. They occur in pairs. The genes in each pair separate (**segregate**) during meiosis when gametes form.*

Gene: A hereditary determinant of a trait. More precisely, a portion of a DNA sequence that codes for one polypeptide.

Allele: One of several possible variants of a gene.

Genotype: All of the genetic traits of an organism, as revealed by breeding experiments.

Phenotype: All of the visible traits of an organism that can be revealed by examining it closely (including microscopically, biochemically, behaviorally, etc.) without any breeding.

Homozygous: Genotype with two identical alleles of the same gene.

Heterozygous: Genotype with two unlike alleles of a particular gene.

Dominant: An allele expressed in the phenotype even when only one such allele is present.

Recessive: An allele expressed in the phenotype only when two such alleles are present. Note: Phenotypes can also be called dominant and recessive.

Blending theory (before Mendel): Early observers examined many traits at once and thus noticed resemblances to both parents. They thought inheritance was like the mixing of fluids; hence, expressions like ''pure blood'' and ''half-blood.''

Particulate theory: Discovered by Mendel in 1865; ignored until rediscovered in 1900.
- Genes come in pairs (in nearly all animals and plants; haploid organisms such as yeast and bacteria may differ).
- Genes come in differing types (dominant and recessive alleles).
- Genes do not blend; they remain discrete.
- Phenotypes are never intermediate. When both alleles occur, only the dominant one is expressed (''law of dominance'').

Important precautions taken by Mendel:
- Used plants of known parentage (derived from pure lines)
- Examined only one trait at a time
- Examined several generations, but examined each separately
- Counted offspring of different types.

Mendel's monohybrid (single-gene) crosses: Mendel chose parents from pure lines, so they were always homozygous. He crossed parents with opposite traits (dominant × recessive). First generation offspring (F_1) all showed dominant trait. Second generation (F_2) had $3:1$ ratio of dominant to recessive phenotypes. Mendel's explanation:
- Use capital A for dominant allele, small a for recessive allele. Parents can thus be represented as AA and aa.
- F_1 are all heterozygous Aa; one gene comes from each parent.
- A and a will separate in the F_1 gametes (**law of segregation**).
- Separate gametes A and a recombine in all possible ways, giving 4 possibilities; ¾ of these (AA, Aa, and aA) have at least one A and will show dominant phenotype, while ¼ will be aa (recessive).

Other single-gene crosses:
- First, determine the genotypes of the parents.
- Each parent produces 1 or 2 types of gametes; combine them to determine F_1.
- Cross $F_1 \times F_1$ if you want to get F_2.

Determine offspring of each cross by a Punnett square:
- List all possible female gametes across the top, and list all possible male gametes in the left-hand column. (The number of possible gametes of each sex will be either 1, 2, 4, or 8, etc., not always the same for both sexes.)
- Fill in each square with the combined genotypes of male and female gametes.
- For traits showing dominance, any genotype with a dominant allele shows dominant phenotype; a genotype with all recessive alleles shows recessive phenotype.

Key 22 Genetics: Crosses with two genes

OVERVIEW *Genes located on different chromosomes separate independently of one another (Mendel's **law of independent assortment**). In Mendel's crosses with 2 genes, all first generation plants were heterozygous for both traits and showed both dominant phenotypes. The F_2 generation showed a $9:3:3:1$ ratio of phenotypes.*

Mendel's dihybrid (2-gene) crosses: Parents chosen from pure lines were homozygous for 2 traits at once. *AABB* × *aabb* gave same results as *AAbb* × *aaBB*. First generation offspring (F_1) were heterozygous for all traits (*AaBb*) and showed all dominant phenotypes. Second generation (F_2) gave $9:3:3:1$ ratio of phenotypes.

Mendel's explanation—the law of independent assortment: Alleles *A* and *a* separate independently of *B* and *b*. Four types of F_1 gametes *(AB, Ab, aB, ab)* are thus produced, in equal proportions. These gametes recombine in all 16 possible ways:

$9/16$ show both dominant phenotypes;

$3/16$ are dominant for the first trait but not the second;

$3/16$ are dominant for the second trait but not the first;

$1/16$ show both recessive phenotypes.

We now know that independent assortment works only if the genes are on separate chromosomes.

Other 2-gene crosses: First, determine the genotypes of the parents. Each parent produces 1, 2, or 4 types of gametes; combine them to determine F_1. Cross F_1 × F_1 to get the F_2. Determine offspring of each cross by a Punnett square (Key 21).

Three-gene crosses: Mendel also studied 3-gene crosses using the methods described above. Parents were *AABBCC* × *aabbcc* (or *AABBcc* × *aabbCC*, etc.). F_1 were all *AaBbCc*. F_2 phenotype ratios were $27:9:9:9:3:3:3:1$. Up to 8 different types of gametes may result in 3-gene crosses.

Key 23   Genetics: Chromosomes and linkage

OVERVIEW   *Chromosome types are distinguished by centromere location. Genes located on the same chromosome tend to be inherited as a group; these groups can, however, be broken apart by crossing over.*

Chromosome morphology: Eucaryotic chromosomes contain proteins (**histones** and **nonhistones**) as well as DNA. Some DNA winds around clumps of histone to form **nucleosomes**, protecting that DNA from transcription. Chromosome **arms** attach to a dense region called the **centromere**. Chromosome types are distinguished as follows:
- **Metacentric** (centromere centrally located; two long arms)
- **Submetacentric** (centromere off-center; arms a bit unequal)
- **Telocentric** (centromere near one end; second arm short)
- **Acrocentric** (centromere at one end; no second arm)

Chromosomal theory of inheritance: Genes are located on chromosomes. The position of each gene is called its **locus**.

Linkage: Genes on the same chromosome are **linked** (unless they are very far apart) and are inherited together as a **linkage group**. The number of linkage groups always equals the haploid number (N).

Crossing over: Sometimes, in prophase I of meiosis, chromosome arms may break and rejoin in reverse order, rearranging linked genes.

Chromosome maps: The probability of crossing over between linked genes varies with the distance between them; recombination frequencies can thus be used to make **chromosome maps (linkage maps)**.

Proof of the chromosomal theory: Experiments show that crossing over of chromosomes always accompanies the recombination of linked genes.

Genetic crosses involving linked genes:
- *AABB* × *aabb* is no longer the same as *AAbb* × *aaBB*.
- F_1 genotypes are heterozygous at each locus. During formation of F_1 gametes, parental combinations of alleles tend to stay together.
- Each F_1 individual is crossed to an organism recessive for all genes being studied (a **test cross**). Phenotypes in the offspring of such a cross correspond to genotypes among the F_1 gametes.

Key 24 Genetics: Sex-linked
inheritance

OVERVIEW *In most diploid organisms, sex is determined by chromosomes called X and Y. In the most common form of sex determination, females are XX and males are XY. For most sex-linked genes, the Y chromosome acts as a "blank," recessive to all alleles; males therefore display the phenotype of the gene on their single X chromosome without regard to dominance or recessiveness.*

Forms of sex determination:
- Females XX, males XY (most plants and animals, including humans); Y chromosome acts as "blank" for most genes.
- Females XX, males XO (some insects); X has no partner.
- Females WZ, males ZZ (birds, Lepidoptera); reverse of XX-XY system; males have two matching Z chromosomes; females have one Z chromosome and a "blank" W chromosome.
- **Haplodiploidy** in bees and ants; males are haploid for *all* chromosomes (not just sex chromosomes); females are diploid.
- Determined by single genes in bacteria and certain fungi.
- Sex determined by environmental temperature or developmental stage in a few fishes.

Sex-linked traits: Traits carried on the X chromosome. *Examples*: white eyes *(w)* in fruit flies (recessive to red); colorblindness and hemophilia (both recessive) in humans.
- $AA \times AY \longrightarrow$ females AA, males AY, both dominant.
- $AA \times aY \longrightarrow$ females Aa, males AY, both dominant.
- $Aa \times AY \longrightarrow$ 4 equally frequent genotypes: Aa (females, dominant phenotype); AA (females, also dominant); AY (males, dominant phenotype); aY (males, recessive).
- $Aa \times aY \longrightarrow$ 4 equally frequent genotypes: Aa (females, dominant); aa (females, recessive); AY (males, dominant); aY (males, recessive).
- $aa \times AY \longrightarrow$ females Aa (dominant); males aY (recessive).
- $aa \times aY \longrightarrow$ females aa (recessive); males aY (recessive).

Holandric traits: Traits carried on Y chromosome (very rare), occurring in males only, and passed directly from father to son.

Key 25   Genetics: Inherited diseases

OVERVIEW   *Humans have small families but keep medical records and genealogies by which inherited diseases may be studied. Sex-linked recessive traits appear more often in males than females and are generally transmitted to males through female intermediaries (carriers). Recessive autosomal traits appear mainly in the offspring of two heterozygous parents who in some cases are genetically related.*

Studying human genetics:
- Humans are poor subjects for genetics in many ways: Small family sizes make genetic ratios hard to determine. Humans cannot be mated just for experiments, and long generation-time limits the number of generations that can be examined.
- But humans are good subjects in other ways: Thorough medical examinations and good medical records; many good genealogical records (family histories).

Human pedigrees: Males are shown as squares, females as circles. Symbols are empty for "normal" individuals, filled-in for affected individuals. Horizontal lines indicate a marriage tie; lines leading down from a marriage tie lead to offspring in order of birth.

KEY EXAMPLES

Biochemical deficiencies (**"inborn errors of metabolism,"** always recessive): Tay-Sachs disease, phenylketonuria (PKU), albinism, hemophilia (sex-linked), fava-bean disease (G6PD deficiency).

Other single-gene traits: Huntington's chorea (autosomal dominant, appearing at age 40–45), ability to taste PhenylThioCarbamide (PTC) (autosomal dominant), colorblindness (sex-linked recessive), sickle-cell anemia (autosomal recessive), A-B-O blood groups (autosomal; A and B dominant; O recessive).

Chromosomal abnormalities: Down's syndrome (extra chromosome, or **trisomy**, of pair #21); Turner's syndrome (XO, a sterile female with only one X chromosome); Kleinfelter's syndrome (XXY, a thin, sterile male with an extra X chromosome).

Patterns of inheritance for rare traits (i.e., traits for which most people are homozygous "normal"):

- **Dominant autosomal traits:** Assume that all individuals showing the trait are heterozygous; homozygous individuals can only come from parents who both show the trait (thus extremely rare).

- **Recessive autosomal traits:** Individuals showing the trait are always homozygous; their parents and children are heterozygous (unless they show the trait also). If both parents show the trait, all their children will, too. If one parent shows the trait, and the other is homozygous normal, then all children are heterozygous. If one parent shows the trait and one is heterozygous, then half the children will show the trait and the others will be heterozygous. The trait cannot appear unless heterozygotes marry, so these traits show up more often if there are **consanguineous** marriages (between genetically related persons).

- **Dominant sex-linked traits:** Sex-linked genes are always expressed in males. Females showing the trait are usually heterozygous.

- **Recessive sex-linked traits:** Sex-linked genes are always expressed in males. Affected males always outnumber affected females in family trees. Females never show the trait unless they are homozygous; such females are rare and occur only if the father shows the trait. Except in such rare cases, mothers and daughters of affected males are unaffected heterozygous individuals (**carriers**). Males with the trait do not have affected fathers or sons, but they often have remote male relatives (like grandfathers, great-grandfathers, or grandsons) with the trait, and the connecting individuals in the in-between generations are always female.

Key 26 Genetics: Other phenomena

OVERVIEW   *Genes may exist in more than two alternative states (**multiple alleles**), or may show varying degrees of dominance. One gene may be expressed in many phenotypic traits. Genes may also interact with each other and with the environment to produce phenotypes.*

Multiple alleles: Genes that have more than two alternative states.

KEY EXAMPLE: A-B-O BLOOD GROUP SYSTEM

In this system, *A* and *B* are both dominant to *o*. When both *A* and *B* occur, both are expressed (they are **codominant**).

BLOOD TYPE	Antigens	Antibodies	Can give blood to:	Can receive blood from:	Possible genotypes:
A	A	anti-B	A, AB	A, O	*AA, Ao*
B	B	anti-A	B, AB	B, O	*BB, Bo*
AB	A and B	neither	AB	A, B, AB, O	*AB*
O	neither	anti-A and anti-B	A, B, AB, O	O	*oo*

Incomplete dominance (incorrectly called "blending"): Some traits show an intermediate condition in heterozygotes. *Example*: in snapdragons, *RR* flowers are red, *Rr* are pink, and *rr* are white.

Dominance can exist in varying degrees:
- None (heterozygotes are midway between homozygotes)
- Partial (heterozygotes closer to one homozygous type)
- Complete (heterozygotes identical to a homozygous type)
- Overdominance ("hybrid vigor") (heterozygotes more extreme than the nearest homozygous phenotype)

Pleiotropy: Condition in which one gene affects many traits.

Gene-gene interactions (polygeny): Several genes combine to make a phenotype. Size is often controlled by **additive polygenes** (each gene adds a small amount; the effects add up to make a bell curve).

Epistasis: One gene "masks" or suppresses the phenotypic expression of a totally different gene (not just a different allele).

Key 27   DNA and the "Central Dogma"

OVERVIEW   *DNA molecules, made of **nucleotides**, are shaped like a **double helix**. Hereditary information flows from DNA to RNA to protein.*

Nucleic acids: DeoxyriboNucleic Acid (DNA) and **RiboNucleic Acid (RNA)** are composed of **nucleotides**, each made of three subunits:
- A phosphate group
- A five-carbon (**pentose**) sugar, either **ribose** or **deoxyribose**
- A nitrogen-containing base, either **adenine (A)**, **guanine (G)**, **cytosine (C)**, **thymine (T)**, or **uracil (U)**. A and G are called **purines**; the others are called **pyrimidines**.

DNA nucleotides: Phosphate, deoxyribose, and A, G, T, or C.

RNA nucleotides: Contain phosphate, ribose, and A, G, C, or U.

Double Helix: DNA contains two strands of nucleotides. Each strand is twisted into a **helix** (corkscrew shape), so the entire molecule is a **double helix**. The two strands are held together by numerous weak **hydrogen bonds** between the nitrogen-containing bases. Hydrogen bonding allows only certain base pairings:
- Adenine pairs only with thymine (**A-T pair** in DNA) or with uracil (in RNA); thymine and uracil pair only with adenine.
- Guanine and cytosine pair only with one another (**C-G pair**).

"Central Dogma": Hereditary information flows from DNA to RNA to protein. Copying DNA to make more DNA is called **replication**. Copying DNA to make messenger RNA is called **transcription**. **Translation** uses mRNA to control the synthesis of a protein sequence. DNA, mRNA, and protein sequences are all **colinear**—they have corresponding codons or amino acids in the same linear order.

Replication: Two DNA strands unravel a bit to make a Y-shaped **replication fork**. Each strand helps synthesize a matching strand, so each new molecule is half newly synthesized and half conserved (**semi-conservative replication**). Nucleotide triphosphates (ATP = Adenosine TriPhosphate, GTP, CTP, and TTP) use some of their stored energy to add a new nucleotide with either A, G, C, or T. ATP can only add A opposite T; GTP can only add G opposite C; CTP can only add C opposite G; and TTP can only add T opposite A.

Key 28 RNA and transcription

OVERVIEW *The three major types of RNA are ribo-somal RNA (**rRNA**), messenger RNA (**mRNA**), and transfer RNA (**tRNA**). All are synthesized in the nucleus; tRNA and mRNA exit via nuclear pores into the cytoplasm.*

RNA differences from DNA: RNA has ribose sugar instead of deoxy-ribose. RNA has bases A, G, C, and U. Uracil occurs instead of thymine; it pairs with adenine. Most RNA is single-stranded (with exceptions in a few viruses only).

Ribosomal RNA (rRNA): In the ribosomes, synthesized in the nucleolus.

Messenger RNA (mMRA): Synthesized in the nucleus by transcription, shaped as a long single strand. It passes into the cytoplasm, where it acts with ribosomes to direct protein synthesis (Key 29).

Transfer RNA (tRNA): In the cytoplasm, used in protein synthesis (Key 29), containing A, G, C, and U, plus some unusual bases.
- Portions of the tRNA molecule can base-pair with other portions to form 3 **loops** (and sometimes a short "extra" loop); a flattened drawing of these loops resembles a clover leaf.
- The true 3-dimensional shape is more complex and L-shaped.
- Over 40 tRNA molecules are known. Each attaches at one end to a specific amino acid. Each also contains a unique 3-base **anticodon** sequence on one of the loops.

Transcription: The copying of a DNA strand to make mRNA, which can occur only if DNA unravels to expose the active strand. After transcription, **excision** removes some intervening sequences (**introns**) that will not be translated into protein, leaving a shorter sequence called an **exon**.

Key 29 Protein synthesis and the genetic code

OVERVIEW *DNA sequences are transcribed into mRNA sequences, then translated into protein sequences. tRNA serves as an adaptor molecule between mRNA codons and amino acids in a protein sequence.*

Translation: Uses an RNA message to help build a protein sequence.
- Coding units in mRNA are called **codons**; each codon contains exactly three bases **(triplet code)**.
- Each codon of mRNA matches with a particular **anticodon** on tRNA. The tRNA thus serves as an "adaptor," controlling the match-up between codons and amino acids.

The genetic code: A set of rules for translating codons into amino acids.
- The code is **commaless**, meaning that nothing separates one codon from the next. **Frameshift mutations**, in which a base is deleted or an extra base inserted, make the bulk of the genetic message unreadable, showing that the message is commaless:

KEY EXAMPLE

Message of triplets: THE CAT ATE THE RAT AND THE HAT
Frameshift with commas: TH CAT ATE THE RAT AND THE HAT
Same change if commaless: THC ATA TET HER ATA NDT HEH AT

- The code is **degenerate** or **redundant**, meaning that different codons may code for the same amino acid. *Example*: CGA, CGG, CGC, and CGU all code for arginine.
- There are three **chain-terminating** or **"stop" codons**: UAA, UAG, and UGA.
- The genetic code is **universal**: all organisms tested use the same code.

Key 30 Mutations

OVERVIEW *Mutations are sudden, permanent changes in DNA. Single-gene mutations include base substitutions, insertions, and deletions. Chromosomal mutations include inversions, translocations, deletions, and duplications. Changes in chromosome number include extra chromosomes, chromosome deficiencies, and polyploidy.*

Mutations: Sudden, lasting changes in hereditary material. Many mutations are caused by thymine-thymine dimers, by mis-matching of nitrogen bases, or by chromosome breakage.

Mutation rates: Natural mutation rates are low (about 1 in a million), but they vary from point to point along a chromosome. X-rays or chemical **mutagens**, like dimethyl sulfoxide, can increase these rates.

Types of single-gene mutations:
- Base-pair substitutions. Chemically, most mutations are **transitions**, substitution of one pyrimidine for another, or one purine for another.
  1. Same-sense mutations (new codon codes for same amino acid)
  2. Mis-sense mutations (new codon codes for different amino acid)
  3. Nonsense mutations (new codon is a "stop" codon)
- Frame-shift mutations, resulting in mis-reading of many codons:
  1. Insertions (extra base added)
  2. Deletions (one or more bases removed)

Types of chromosomal aberrations:
- **Inversions** (part of a chromosome reversed end-to-end)
- **Translocations** (chromosome fragment joins a different chromosome)
- **Duplications** (part of a chromosome occurs twice)
- **Deletions** (part of a chromosome is lost)

Changes in chromosome number:
- **Aneuploidy** (loss or gain of one chromosome at a time):
  1. **Monosomy** (one chromosome lost from a diploid pair)
  2. **Trisomy** (an extra chromosome added to a diploid pair)
- **Polyploidy** (addition of entire haploid sets). Levels of polyploidy include tetraploid (4N), hexaploid (6N), and octaploid (8N).

Key 31   Recombinant DNA
technology

OVERVIEW   *Recombinant DNA (genetic engineering)*
techniques manipulate DNA fragments and splice them into
combinations that do not occur naturally.

Excision of DNA sequences: **Exonucleases** attack the ends of DNA
molecules; **endonucleases** attack interior DNA sequences. Of these,
restriction endonucleases (**restriction enzymes** such as **EcoRI,
BamHI, HindIII, BalI**) attack only particular sequences within
DNA molecules (usually **palindromes**, sequences that read the same
in both directions). Most restriction endonucleases cut the two DNA
strands in slightly different locations, forming **cohesive ("sticky")
ends** that can join with one another or with other cuts made by the
same enzyme. **Blunt** or **flush ends** are less useful.

Insertion of DNA sequences: DNA sequences with "sticky" ends will
join by themselves if they are cut with the same endonuclease and
thus have matching ends. "Sticky" ends can join in any order, so
there must be a way of detecting only the desired sequences. The
enzyme **DNA polymerase** is used to fill in gaps in the sequence, then
another enzyme, **DNA ligase**, seals the joined molecules.

DNA vectors (cloning vehicles): DNA sequences that replicate easily.
Bacterial **plasmids** (Key 69) and some viruses (like M13) are naturally occurring DNA vectors. Most DNA vectors are circularly
closed, not linear. Ethidium bromide, which binds more easily to
linear DNA, can isolate vectors.

Cloning of DNA sequences: A **DNA probe** (a vector bearing an artificially inserted sequence) is inserted into a cell, and the cell is allowed
to replicate. The vector must contain genetic traits that permit selection of those cells that have incorporated the vector.

Applications of genetic engineering:
- **Investigative:** Making nucleic acid sequences for research
- **Commercial:** Growing useful genes and gene products, like
  human hormones or pharmaceuticals grown in bacteria
- **Agricultural:** Adding essential amino acids to plant proteins, or
  allowing plants to fix atmospheric nitrogen without fertilizers.
- **Medical:** Inserting genes someday to cure genetic defects.

Key 32 Viruses

OVERVIEW *Viruses are fragments of nucleic acid, often surrounded by protein, that can replicate only with the aid of intact cells. All viruses have a **lytic cycle**, in which they invade a cell, replicate inside, then rupture the cell and release their progeny. Some viruses also have a **lysogenic** cycle, in which they lie dormant and replicate as part of the host DNA. Viruses are classified by the type of nucleic acid (either DNA or RNA).*

General characteristics: Viruses are fragments of nucleic acid (DNA or RNA, never both) often surrounded by protein. A few viruses also have capsules derived from host membranes. Viral shapes can be helical, icosahedral (20-sided), or complex (with head and tail). Viruses have very few of the properties of life (Key 1). They cannot reproduce without the gene-replicating machinery of the host cell.

Lytic cycle (in all viruses):
1. Virus first attaches to host cell and injects nucleic acid only.
2. Viral DNA or RNA is replicated using the host cell's enzymes.
3. Host cell ruptures, releasing thousands of new virus particles.

Lysogenic cycle (in some viruses only):
1. Viral DNA inserts itself into the host cell chromosome.
2. Virus then hides (**lysogenic stage**), replicating as part of host DNA.
3. Upon ''activation,'' the virus takes over the cell's reproductive machinery and resumes the lytic cycle.

KEY TYPES OF VIRUSES

DNA viruses
  With single-stranded DNA:
    Parvoviruses
    ϕX174 bacteriophage
  With double-stranded DNA:
    T2, T4, T6 bacteriophages
    Lambda bacteriophage
    Adenoviruses
    Herpes viruses
    Pox viruses (smallpox, etc.)

RNA viruses
  With single-stranded RNA:
    PicoRNA viruses (polio, etc.)
    Togaviruses (rubella, etc.)
    Rhabdoviruses (rabies)
    Measles and mumps viruses
    Influenza viruses
    Retroviruses (AIDS, etc.)
    Tobacco mosaic virus (TMV)
  With double-stranded RNA:
    Reoviruses

Key 33   Embryology: Normal development

OVERVIEW   *Most animals go through a series of common embryological stages, including a ball-shaped **blastula** and a **gastrula** stage during which **germ layers (endoderm, mesoderm, ectoderm)** are formed. **Neurulation** (formation of the nervous system) follows in vertebrate embryos.*

Ontogeny: Embryonic development is only part of **ontogeny**, a life-long process of prenatal development, birth, maturation, adulthood, and senility. An organism in its early developmental stages is called an **embryo**; embryonic stages are similar in many organisms. A developing mammal is called a **fetus** from organ formation to birth.

Development in amphibians:
- **Early stages:** The fertilized egg **(zygote)** contains stored food **(yolk)**. The first cleavage in most animals runs vertically, from pole to pole. The second cleavage is usually vertical, at right angles to the first. The third cleavage is horizontal. Cells of upper half (animal hemisphere) contain less yolk and divide faster than yolk-rich lower half (vegetal hemisphere).
- **Blastula:** A hollow ball of cells with a cavity **(blastocoel)**. All animals go through a blastula stage.
- **Gastrulation:** Cells begin to **invaginate** (tuck in) along the **dorsal lip** to form a new cavity, the **archenteron** (''primitive gut''). The entrance to the archenteron is called the **blastopore**.
- **Neurulation:** Nervous system forms from a pair of folds which close over. Mesodermal **somites** (muscle blocks) form at about this time.
- Development of sense organs, gill slits, and circulatory system.
- Hatching into gill-breathing tadpole stage.
- **Metamorphosis** into an adult (controlled by thyroid hormone).

Germ layers (usually formed during gastrulation):
- **Ectoderm:** forms outer layer of the body (epidermis of skin, skin glands, hair, etc.); also nervous system (brain, nerves, etc.).
- **Mesoderm:** forms most muscles and bones, plus entire circulatory system, excretory system, reproductive system, and many other organs.
- **Endoderm:** forms the inner lining of the gut, plus the lungs, trachea, middle ear, liver, and pancreas.

Key 34   Embryology: Control of development

OVERVIEW   *Many developmental changes are controlled by "organizers," which induce changes in cell growth. Experiments on amphibians show that the dorsal lip is an organizer for gastrulation movements and that cells on the roof of the archenteron are an organizer for the formation of the nervous system. Many organizers contain RNA. Hormones may control developmental processes like metamorphosis.*

Organizer theory: Organizers are substances secreted by cells that induce changes in the growth of other cells. They were first studied by transplant experiments on amphibians.
- Many organizers contain RNA. Chemical extracts of organizers are effective inducers, but **ribonuclease (RNAse)** abolishes this ability.
- Various mutational agents **(mutagens)** and cancer-causing agents **(carcinogens)** are also effective inducers.

Dorsal lip (Key 33): An organizer for gastrulation. Dorsal lip transplanted to another embryo induces gastrulation in a second place and results in a "double gastrula."

Notochord (roof of the archenteron): An organizer for the formation of the nervous system.
- If normal gastrulation is prevented, no nervous system forms. (Stripping the jelly-like capsule from eggs prevents the dorsal lip from tucking in; neural tube never forms.)
- "Double gastrula" (see above) develops two neural tubes.
- Notochord transplanted to another embryo induces a second nervous system to form; transplants between species show that induced nervous systems contain no cells from the transplant.
- Cell-free extract of notochord can induce a second neural tube.

Role of hormones: The insect hormone **ecdysone** causes chromosome puffing, which exposes DNA sequences needed for metamorphosis and wing development. In larval amphibians (tadpoles), **thyroxin** causes metamorphosis into an adult. Destroying thyroxin prevents metamorphosis; giving thyroxin early induces premature metamorphosis.

Key 35   Gene control in development

OVERVIEW   *In general, genes control development by making proteins, including enzymes, that regulate other processes. These proteins need to differ with cell type, sex, developmental stage, and physiological condition. The operon theory provides one model to explain many phenomena of gene regulation in procaryotes.*

Gene control: Genes control cell processes by making enzymes and other proteins. However, these proteins need to vary from one cell type to another. *Example*: All human cells have genes for insulin, but only the pancreas cells make insulin, and only when it is needed.
- Different proteins are needed in embryos, children, and adults.
- Many genes or gene products (for beard development, antlers, or milk production) are needed in one sex but not the other.
- Gene products are also needed in differing amounts. When certain substrates are present, more enzymes are needed to break them down, but it is wasteful to produce more than is needed.
- All genes were once thought to make enzymes (**"one gene, one enzyme"**). Now we know that some gene products regulate other genes.

Types of gene control:
- **Positive control:** genes normally "off"; turned "on" when needed.
- **Negative control:** genes normally "on"; can be turned "off."
- **Constitutive** (no control): a gene is always "on."
- **Transcriptional control:** regulates whether or not a gene is transcribed.
- **Translational control:** Regulates whether or not mRNA is translated into protein.

Gene control in procaryotes: Procaryotic genes are often regulated as parts of **operons**, groups of genes *regulated together as a unit*, meaning that the genes are all turned on or off at the same time (**coordinate regulation**). The genes in an operon are usually related in function, as by making enzymes that control successive steps in a biochemical pathway. All known operons are transcriptionally regulated.
- The *lac* operon of *E. coli* is the best-known operon. It contains, in order, a **repressor gene** *(lacI)*, **a promoter** *(lacP)*, an **operator** *(lacO),* and three **structural genes** (lacZ, lacY, and lacA) that make enzymes concerned with lactose metabolism.

- Transcription requires that RNA polymerase first bind to the promoter region. The *lac* **mRNA** that results contains all three structural genes.
- The repressor gene makes a **repressor protein**, which binds to the operator, preventing transcription, a form of negative control. Transcription stops before reaching the structural genes.
- When gene products are needed, an **inducer** protein (made elsewhere) combines with the repressor and inactivates it. This process, **derepression**, exposes the operator, allowing transcription to proceed.
- Transcription of *lac* mRNA also requires **cyclic AMP** and **catabolite activator protein (CAP)**, a form of positive control.
- Some operons are **autoregulated**: the gene product acts as its own repressor. A small amount of product, always present, represses the needless synthesis of any more product. If this small amount is used up, the gene is derepressed and more product is made.

Gene control in eucaryotes: Gene control is more complex and less well understood in eucaryotes.
- Most mRNA in eucaryotes contains only one gene at a time.
- Much eucaryotic DNA is never translated, but consists of repeated sequences which occur hundreds to several millions of times.
- Most eucaryotic DNA is bound to **histone** proteins, forming **chromatin**. DNA cannot be transcribed when the histones remain tightly bound; this may be the key to one method of gene regulation.
- Oocytes need genes for ribosomal RNA more numerously to make the ribosomes needed to make yolk proteins. **Gene amplification** greatly increases (temporarily) the number of copies of these genes.
- Many eucaryotic genes belong to **gene families**, believed to have originated from the duplication of one original gene and the subsequent mutation of the resultant copies.
- Genes that produce antibodies have a special mechanism that pieces together several gene fragments, allowing a few hundred fragments rapidly to generate many millions of new variants.

Theme 6   STRUCTURE AND FUNCTION OF BODY SYSTEMS

*E*very organism has some means of carrying out each of its life functions; complex organisms usually have specialized organs for each of their major functions. Every organism must have: (1) a way of obtaining food and extracting nutrients from it, (2) a way of exchanging gasses with its surroundings, (3) a way of transporting and circulating materials within its body, (4) a way of controlling water balance and getting rid of wastes, and (5) a way of reproducing. Many animals also have specialized cells for warding off infections, specialized skeletal tissue for support, and special contractile tissue called muscle.

Key 36   Nutrition and transport in plants

OVERVIEW   *Leaves are the principal organs of photosynthesis in most plants. The light reactions of photosynthesis take place most efficiently in the **palisade layer**, and the dark reactions in the **spongy mesophyll**. Vascular tissues (**xylem** and **phloem**) are responsible for transport of materials within the plant.*

Leaves: Made up of:
- **Upper epidermis:** coated with a waxy **cuticle**
- **Palisade layer:** containing the highest chloroplast density—light reactions of photosynthesis (Key 17) most efficient here.
- **Spongy mesophyll:** dark reactions of photosynthesis (Key 17) most efficient here because air spaces facilitate gas exchange.
- **Veins (extensions of xylem and phloem):** run through the spongy mesophyll layer.
- **Lower epidermis:** lined with a waxy cuticle; contains pores for gas exchange called **stomates** or **stomata** (Key 38).

Phloem transport: The phloem transports photosynthetic products from leaves to other parts of the plant, principally downward through the stem. The principal transport cells in phloem are **sieve tube** cells.

Xylem and transpiration: Water and dissolved minerals (ions, including K^+, Mg^{+2}, Ca^{+2}, NO_3^-, PO_4^{-3}) ascend from roots through stems to upper parts of the plant, traveling through tube-like **tracheids** of the xylem. Loss of water from leaves is called **transpiration**. The ascent of sap seems to be governed largely by **transpiration pull**, or reduced pressure from above, a process requiring long, unbroken columns of fluid with no air bubbles. Root pressure also helps a bit.

Key 37   Nutrition in animals

OVERVIEW   *Unicellular organisms ingest food by phagocytosis. Most digestion in lower animals is intracellular, but extracellular digestion (both mechanical and chemical) predominates among higher animals. The "assembly line" (mouth-to-anus) digestion allows regional specialization to take place, so that digestion happens in stages.*

Intracellular digestion: Predominates in lower animals. In **phagocytosis,** folds of the cell membrane engulf some food material originally outside the cell, forming a vacuole. Chemical digestion then follows when these vacuoles fuse with lysosomes (Key 12).

Extracellular digestion: Predominates in higher animals. Includes both **mechanical digestion**, in which food is minced or crushed, exposing more surface area, and **chemical digestion**, in which food is broken down chemically with the help of enzymes.

Mouth: Major site of mechanical digestion (using teeth, etc.). Chemical digestion begins with an **amylase** enzyme in saliva that breaks down starches. Food passes from mouth to stomach via the **esophagus.**

Stomach: Major site of protein digestion. **Pepsin** breaks proteins into small peptides. Hydrochloric acid (HC1) acidifies stomach contents; this helps pepsin, which works best at acidic pH (near 2.0). Mechanical digestion also occurs by contraction of three muscle layers, kneading food. Some species (like birds) have a storage pouch **(crop)**, followed by a strongly muscular **gizzard**, specialized for mechanical digestion.

Liver: Secretes **bile**, containing bile pigments (from hemoglobin) and bile salts. Bile is stored in the **gall bladder** until needed.

Small intestine: Long and coiled, with large surface area. Includes:
- **Duodenum:** digestion of lipids, carbohydrates, and proteins, using enzymes secreted by duodenum itself and by **pancreas.** Lipids are **emulsified** by **bile salts** secreted by the liver.
- **Jejunum:** chemical digestion of most nutrients completed here.
- **Ileum:** site of most **absorption** of digestion end-products.

Large (large-diameter) intestine: Includes **caecum** (size varies; site of microbial digestion of cellulose in herbivores) and **colon** (site of much water absorption).

Anus: Undigested wastes are eliminated.

Key 38 Gas exchange

OVERVIEW *Flatworms and other flattened animals need no special organs for gas exchange because no cell is very far from a body surface. More complex animals use lungs, gills, or tracheal tubes. Plants rely on special holes* (*stomates*) *in the lower epidermis of each leaf.*

Anatomy in land vertebrates: Nostrils take air into **nasal cavity**, then into **pharynx**. Floor of pharynx opens behind mouth into **larynx** (voice-box); entrance to larynx is guarded by **epiglottis**. The **trachea, bronchi**, and **bronchioles** form tree-like branchings within each lung. The **lungs** have **air sacs** lined with box-like **alveoli**.

Air exchange: In **inhalation (inspiration)**, diaphragm contracts and moves downward while intercostal muscles raise rib cage. In **exhalation (expiration)**, muscles relax, rib cage falls, diaphragm springs upward.

Gas exchange in alveoli: Oxygen enters capillaries of lung through thin walls; CO_2 leaves capillaries and diffuses into air sac.

Gas exchange within capillary blood: In lung alveoli, oxygen enters red blood cells, combines with hemoglobin, and is transported as **$KHbO_2$ (oxyhemoglobin)**; bicarbonate ions enter blood cells and are split into water and CO_2. The reverse occurs in body tissues: oxyhemoglobin breaks down to release oxygen; CO_2 and water combine to form bicarbonate ions (HCO_3^-).

Gill systems: In fishes and many other aquatic animals, thin-walled arteries run through gills, with direction of blood flow usually opposite to flow of water (**counter-current** exchange). Oxygen diffuses into these arteries; CO_2 diffuses out into water.

Insect tracheal systems: Air diffuses through numerous branched tubes (**tracheae**). Rhythmic muscular contractions force air in and out when flying, but air movement is passive most of the time.

Stomates in plants: Openings in the lower epidermis of each leaf. Two large kidney-shaped **guard cells** flank the stomate on either side. Swelling of guard cells closes the stomate; shrinking of guard cells opens it. Gasses enter and leave by passive diffusion. During photosynthesis (by day), CO_2 enters and oxygen diffuses out. During respiration (at night), CO_2 and water diffuse out; O_2 enters. Stomates usually close when water loss would be excessive.

Key 39   Internal transport and circulation

OVERVIEW   *Very small or very thin organisms need no special system for internal transport. Many invertebrates have an **open** system, with blood vessels opening into a general circulatory cavity or **hemocoel**. Vertebrates have a **closed** circulatory system: their hearts pump blood from **atrium** to **ventricle** and then through the major arteries; veins return blood to the heart.*

Simple forms of transport:
- **Cytoplasmic streaming (cyclosis):** Cytoplasm in all eucaryotic cells continually flows and changes direction.
- **Diffusion:** Passive transport in all organisms, effective only at distances of a few cells. This may suffice for organisms in which each part is only a few cells away from a body surface, but larger organisms need circulatory systems.

Internal transport in plants: Uses xylem and phloem (Key 36).

Open circulatory systems: System in which a body cavity or **hemocoel** contains most of the circulating fluid, as in insects.
- The pumping action of a **heart** drives fluid forward through an **aorta**, then into a series of **arteries**. No veins exist; used blood seeps into **sinuses** that drain into the hemocoel.

Closed circulatory systems: System in which blood is everywhere contained within a vessel, as in all vertebrates.
- The heart may have 2 to 4 chambers. The heartbeat originates from a **pacemaker** at the **sinoatrial node**. Highest pressure, at maximum contraction, is called **systole**; lowest pressure is called **diastole**. In mammals, the **right atrium** and **ventricle** pump oxygen-poor blood to the lungs; the **left atrium** and **ventricle** pump oxygen-rich blood from the lungs to the body through the **aorta**.
- **Arteries** carry blood from the heart to the body's tissues.
- **Veins** return the blood from the body's tissues back to the heart. Valves in veins prevent blood from flowing backward.
- Vertebrate blood is always red because of the oxygen-carrying pigment **hemoglobin**, carried in **red blood cells (erythrocytes)**.

Key 40 Osmoregulation and excretion

OVERVIEW *Freshwater organisms tend to gain water across membrane surfaces and must actively get rid of it. Land and marine organisms tend to lose water; they must retain water and excrete salt. Vertebrate kidneys filter the blood first, then resorb useful molecules.*

Osmotic pressure: Measures the level of dissolved ions in solution.

Hypotonic solutions (low osmotic pressure, few dissolved ions): Cells swell (or may burst) because water diffuses in. Freshwater organisms always gain water from hypotonic surroundings; they void lots of dilute urine and may actively take up the ions.

Hypertonic solutions (high osmotic pressure, many ions): Water diffuses out; cells shrink. Marine and land animals lose water across membranes. They excrete concentrated urine or salt-rich fluids.

Isotonic solutions: Cells have the same concentration of dissolved ions. Water enters and leaves at the same rate; cells stay the same size.

Simple excretory systems: Certain freshwater protists pump water out by **contractile vacuoles**. Many small, aquatic animals allow wastes to diffuse out. Flatworms have single-celled excretory tubules called **flame cells** (Key 80).

Nephridial systems: Tubules (**nephridia**) drain coelomic fluid from the body cavity and exchange ions with small blood vessels nearby.

Vertebrate kidneys: Cortex (outer layer) includes mostly glomeruli and convoluted tubules. **Medulla** (inner layer) is made of several **medullary pyramids**, which contain Henle's loops.

Kidney tubules: Blood plasma is filtered from a set of thin-walled blood vessels (the **glomerulus**) into **Bowman's capsule**. In the **proximal convoluted tubule**, the blood resorbs glucose and some ions. In mammals, **Henle's loop** resorbs water. In the **distal convoluted tubule**, more ions return to the blood. **Collecting tubules** finally concentrate the urine and drain into the **renal pelvis**, which drains into the **ureters**.

Nitrogen wastes: In mammals, the principal nitrogen waste is **urea**, but some organisms secrete uric acid or ammonium salts instead.

Other organs of excretion: Lungs and gills get rid of CO_2. Animals excrete salt and nitrogen wastes through the skin.

Key 41 Reproductive organs

OVERVIEW *Females have ovaries in which eggs form. The eggs travel through an oviduct into the uterus or uteri. Oviparous females then lay eggs, while viviparous females bear their young alive. Sperm are produced in the testes of males and travel through the vas deferens.*

Asexual reproduction: Reproduction without genetic recombination.

Sexual reproduction: Reproduction including genetic recombination.

Gonochorism: Male and female sex organs located in separate organisms.

Hermaphroditism: Both male and female sex organs in same individual.

Oviparous: Egg-laying. Eggs usually **fertilized externally**, but reptiles and birds have **internal fertilization**, requiring copulatory organs.

Viviparous: Giving birth to live young; fertilization always internal.

Plant reproduction: see Keys 72 and 78.

Female reproductive organs in vertebrates:
Ovaries: Female gonads, produce eggs (ova), also estrogen (Key 51).

Oviducts (Fallopian tubes): Carry eggs from ovaries to uteri. In birds, a **shell gland** surrounds eggs with protective coatings.

Uterus: Womb where embryo grows. Made of 3 layers: **epimetrium** (thin connective tissue capsule); **myometrium** (thick, muscular layer); **endometrium**, a sensitive, mucus-coated, inner layer with periodic (monthly) fluctuations in thickness (ending in menstruation) under hormonal control. Shape may be **duplex** (two separate, paired uteri, as in nonmammalian vertebrates), **bipartite** (two separate uteri emptying into a common **cervix**, as in many hoofed mammals), **bicornuate** (Y-shaped, with two long **horns** and a common **body** and **cervix**, as in most carnivores), or **simplex** (a single **body** and a neck-like **cervix**, as in primates).

Vagina: Tubular sheath whose lining secretes a lubricating mucus.

External genitalia:
* **Labia majora:** Outer folds, dry on outside.

- **Labia minora:** (Correspond to shaft of male penis); sensitive inner folds containing erectile tissue **(corpora cavernosa)**.
- **Clitoris:** (Corresponds to tip of penis); sensitive erectile body located where labia minora come together; contains erectile tissue **(corpus spongiosum)**.
- **Hymen:** Delicate membrane guarding entrance to vagina.
- **Bartholin's glands:** Just outside vaginal entrance, secrete a lubricant that sometimes contains a sex-attractant **pheromone**.

Male reproductive organs in vertebrates:

Testes: Male gonads; contained in **scrotal sacs** in mammals; correspond to the labia majora of females. Seminiferous tubules produce sperm; interstitial cells produce testosterone (Key 51).

Efferent ductules: Carry sperm from testes to epididymis.

Epididymis: A coiled tube, secretes much of the seminal fluid.

Vas deferens (ductus deferens): Carries sperm and seminal fluid from testes to penis.

Seminal vesicles: Secrete a thick, nutritive fluid containing fructose, which provides energy for the swimming motons of the sperm.

Prostate gland: Secretes a thick, milky fluid containing enzymes that help liquefy the seminal fluid after ejaculation.

Cowper's (bulbourethral) gland: Secretes a clear fluid that dilutes the seminal fluid and lubricates the urethra.

Male urethra: Carries both urine and semen; made of 3 portions (prostatic, spongy, penile).

Penis: Organ of copulation, corresponding to clitoris + labia minora of female. Contains three bodies of erectile tissue: one **corpus spongiosum (corpus cavernosum urethrae)** and two **corpora cavernosa (corpora cavernosa penis)**. All three become rigidly engorged with blood under sexual stimulation. No penis exists in species where fertilization is external.

Key 42   Immunology

OVERVIEW   *Many animals have a special group of cells that respond to infections by a variety of defenses. Antibody-mediated immunity involves a group of proteins called* **immunoglobins**, *which combine with foreign substances* (**antigens**) *to inactivate the invaders. In cell-mediated immunity, phagocytic cells are activated in such a way that they seek out and engulf the infected host cells.*

Immune response: A defense system based on special cells that respond to infection by increasing in number and fighting the infection. All vertebrates (and a few invertebrates) show immune responses; the immune responses of mammals are the most elaborate.

Antigen: Any substance capable of evoking an immune response. Immune responses are very **specific** to particular antigens.

Primary response: The response to an initial encounter is slow, and only moderate numbers of defensive cells are recruited.

Secondary response: Subsequent encounters with the same antigen produce a much more rapid response and a much stronger reaction using many more defensive cells.

Memory: The immune system recognizes specific antigens that it has encountered before, so primary and secondary responses differ.

Antibody-Mediated Immunity (AMI): Antibody-mediated immunity is most effective against bacterial infections. Certain white blood cells (**B-lymphocytes or B-cells**) produce special infection-fighting proteins called **antibodies** or **immunoglobins (Ig)**. There are several classes of antibodies:
- **Ig-G** (about 85% of total): Y-shaped, with 2 light and 2 heavy chains
- **Ig-M** (about 13% of total)
- Less common types: Ig-A, Ig-D, Ig-E

Antigen-antibody complexes: All antibodies disable antigens (or cells bearing antigens) by combining with them to form insoluble cross-linked complexes or clumps. The antigen-antibody complex also activates **complement** (a group of proteins that activate phagocytosis), inflammation (**histamine reaction**), and rupture (lysis) of cells.

Cell-Mediated Immunity (CMI): Certain antigens activate a response that kills diseased body cells, not just the disease agents. Cell-mediated immunity is most effective against viral and certain eucaryotic infections.

Major histocompatibility complex (MHC): Cell surface proteins that allow the body to recognize its own cells. Cells of the CMI system attack only those cells bearing *both* the MHC and certain specific antigens indicating that those cells have been infected.

Macrophage: Type of white blood cell that engulfs infected cells and kills them by phagocytosis.

T-lymphocytes (T-cells) and their roles:
- **Cytotoxic T-cells:** Attack infected cells by making holes in their cell membranes.
- **Suppressor T-cells:** Turn other lymphocytes off when they are no longer needed.
- **T-helper cells:** Activate macrophages by releasing **lymphokines** such as **interleukin**.

Disorders of the immune system:
- **Autoimmune disorders:** The immune system attacks its own body, as in rheumatoid heart disease, rheumatic fever, systemic lupus, and probably juvenile diabetes.
- **Allergy:** Hypersensitivity to common environmental antigens (**allergens**), like ragweed pollen. Reactions usually involve Ig-E.
- **Immunodeficiency:** Inability to make enough lymphocytes. *Example:* **Acquired Immune Deficiency Syndrome (AIDS)** results from a viral attack upon the T-cells.

Monoclonal antibody technique: Medical technique that makes large numbers of antibodies against a particular antigen.
1. Antigen is injected into mouse, which develops antibodies against it.
2. Spleen is removed from mouse and B-cells isolated from it.
3. B-cells are fused with fast-growing **myeloma** cells from a tumor.
4. Successfully fused (**hybridoma**) cells are grown on a medium that kills other cell types.
5. Hybridoma cells grow rapidly, producing lots of one specific antibody.

Key 43 Skeleton in general

OVERVIEW *Various types of skeletal systems can help support the body's shape. Fluid-filled cavities can form a hydrostatic skeleton. Skeletons can also be built of rigid structures on the outside (exoskeletons) or on the inside (endoskeletons). Skeletal proportions (and therefore body proportions) are often dictated by area/volume relationships.*

Hydrostatic skeleton (as in many worms): A series of fluid-filled body cavities (**coelomic cavities**, see Key 81) can form a skeleton. Such a skeleton can be very functional in burrowing through loose sand or soil: those segments constricted by circular fibers bulge forward and lengthen the body, while segments compressed along the body axis by longitudinal fibers swell sideways and anchor those parts of the body against slipping backward.

Exoskeleton (as in insects and other arthropods): Body parts are hollow, with hardened tissue on the outside. In small organisms (most insects, spiders), exoskeleton can be made of **chitin**, but in larger organisms (lobsters), chitin layer is usually strengthened by calcium salts.
- Muscles are arranged on the inside, spanning joints. Most muscles span one joint at a time.
- The weight of the exoskeleton limits its size, especially on land.
- Growth is a problem once the skeleton hardens, so animals with exoskeletons **molt** at certain intervals: the animal cracks through and sheds its skeleton, exposes a new skeletal layer underneath, inflates its body, and allows its new, larger skeleton to harden.

Endoskeleton: A skeleton on the inside, as in vertebrates. Muscles are arranged surrounding the bones of the skeleton. Animals with endoskeletons are generally larger than those with exoskeletons. Growth is less of a problem because skin can remain flexible.

Area-volume relationships: Skeletal proportions are often dictated by area/volume relationships, which limit body shapes and sizes.
- Mass and muscle strength are both related to volume.
- Efficiency of respiration, digestion, or heat exchange are related to surface area, so larger animals need more heavily folded surfaces.
- Strength of supporting legs is related to cross-section area, while the weight needing support is related to volume. Thus, large animals need disproportionately thicker, cylindrical, and more vertical legs.

Key 44 Vertebrate skeleton

OVERVIEW *The skeleton contains both bone and cartilage tissues. Most embryonic cartilage later turns to bone.*

Cartilage tissue: Not as strong as bone, but resists shock better. Smoother surface is better at moving joints. No blood supply; cells nourished by diffusion only. Grows much more rapidly than bone, until diffusion cannot meet nutritional needs; then cells die or are replaced by bone. Types of cartilage are: **hyaline cartilage** (simple matrix); **fibrous cartilage** (collagen fibers added); **elastic cartilage** (elastic fibers added); **calcified cartilage** (calcium salts added).

Bone tissue: Stronger than cartilage, but less resistant to shock and grows more slowly. Internal blood supply nourishes interior and keeps cells alive, allowing constant restructuring, repair, and healing of injuries.

Compact (lamellar) bone tissue: Made of **Haversian systems** (concentric cylinders surrounding a central Haversian artery). As bones restructure, new Haversian systems align parallel to greatest stress.

Spongy (cancellous) bone tissue: Built of struts **(trabeculae)** separated by spaces containing blood and/or marrow.

Bone growth: There are two kinds of bone formation **(ossification)**: **Dermal** or **intramembranous ossification**, within dermis of the skin, forms **membrane bones** (clavicle, bones of skull roof, palate).

Endochondral ossification: Pre-shaped **(preformed)** as fast-growing cartilage tissue; bone then replaces the cartilage.
- **Visceral bones** (derived from gill arches): Alisphenoid; malleus, incus, stapes; hyoid apparatus; cartilages of larynx and trachea.
- **Somatic bones** (derived from embryonic somites). Includes:
1. **Axial** (derived directly from somites): Braincase (ethmoid, sphenoid, occipital series); vertebrae, ribs, sternum.
2. **Appendicular** (derived via limb bud): Scapula, coracoid, humerus, radius, ulna, bones of wrist and hand; innominate bone, femur, tibia, fibula, bones of ankle and foot.

Growth of long bones: Cartilage first grows in all directions. A **primary ossification (diaphysis)** forms in the middle of the bone, and a **secondary ossification (epiphysis)** forms at either end. An **epiphyseal cartilage** between diaphysis and epiphysis marks the region of fastest growth. The growing diaphysis finally replaces the epiphyseal cartilage; bone growth ceases when **epiphyseal fusion** is complete.

Key 45 Muscles

OVERVIEW *Contraction results from the sliding of thin filaments of actin lengthwise between the thick myosin filaments. See also Key 16.*

Fine structure of striated muscle:
- Protein fibers are perpendicular to alternating light and dark bands.
- Dark **A-bands** are made of **thick filaments** of the protein **myosin**; the protein **troponin** usually binds to myosin to form **tropomyosin**.
- Light **I-bands** are myosin-free, but they contain **thin filaments** of **actin**.
- Midway through each I-band runs a dark **Z-line**; thin filaments attach to this Z-line. The interval from one Z-line to the next is the unit of contraction, called a **sarcomere**.
- Extensions of the plasma membrane (**sarcolemma**) run along each Z-line to form **transverse tubules (T-tubules)**.

Mechanism of contraction (''sliding filament theory''):
- T-tubules supply calcium ions and oxygen to the contracting fiber.
- When a nerve signal arises, the plasma membrane becomes more permeable to calcium.
- Calcium ions rush in and bind to troponin preferentially, releasing myosin.
- Myosin molecules, shaped somewhat like golf clubs, now bind to actin.
- The heads of the myosin molecules now rotate, moving the actin fibers and causing them to slide along the myosin fibers.
- After its head rotates, each myosin fiber detaches from its actin fiber. The head resumes its original position and forms a new attachment with another actin fiber.
- Myosin fibers are staggered from one another; some rotate against their actin fibers while others are releasing their holds and recovering. In this way, muscle contraction is smooth and even instead of jerky.

Theme 7 CONTROL OF BODILY FUNCTIONS

Body functions are controlled by a rapid-response system using nerves and also by hormones that act more slowly. Nerve cells (neurons) conduct impulses along their membrane surfaces and transmit these impulses to adjoining cells by releasing neurotransmitters. The brain and sense organs (eyes, ears, etc.) contain many special types of neurons. Hormones are chemicals transported to their targets by the blood stream. An imbalance in the amount of any hormone results in disease. Most sex hormones are steroids; many other hormones are proteins or protein derivatives.

Certain body processes observe daily rhythms. Many organisms can use the changes in day length to monitor seasonal changes. Animals can modify the circumstances of their lives through behavior, which may be either simple or complex, and either innate or learned.

Key 46   Nervous System: Neurons

OVERVIEW   *Nerve tissues are made of nerve cells (neurons) and other cells (neuroglia). A nerve impulse begins as a depolarization of the normally polarized cell membrane, and travels as a wave of depolarization. The neuroglia cells aid in the structural support and nourishment of the neurons.*

Structure of nerve cells (neurons):
- **Cell body:** Includes both **nucleus** (normal in structure, usually round) and cytoplasmic portion (**perikaryon**, surrounding nucleus). Most of the perikaryon is very rich in grain-like vesicles of rough endoplasmic reticulum (**Nissl granules**), rich in RNA. *Exception*: **axon hillock** (at base of axon) is largely Nissl-free.
- **Dendrites** (often numerous): Branching processes whose membranes conduct nerve impulse toward cell body. Interior cytoplasm resembles cytoplasm of cell body, but with fewer Nissl granules.
- **Axon**: Longer, cylindrical processes, essentially unbranching over most of their length, usually conducting impulses away from the cell body and usually more rapidly than in dendrites. Many axons are surrounded by a **myelin sheath** of fatty **Schwann cells**, providing electrical insulation by wrapping plasma membrane around axon in the manner of a jelly roll. Impulses in myelinated axons travel much faster by **saltatory conduction** (''jumping'') from one constriction (**node of Ranvier**) to the next.

Nature of the nerve impulse:
- In the ''resting'' (non-conducting) nerve cell, a **sodium pump** actively transports Na^+ ions across the plasma membrane to the outside of the cell. Potassium (K^+) ions rush in to neutralize the electrical charge, but they do not quite compensate, so there is a net negative charge of about 60 millivolts (-60mV) inside the cell.
- Changes that **hyperpolarize** the membrane, or that depolarize it only slightly, travel a short distance along the membrane before they decay. They may locally alter the probability of an action potential.
- **Neurotransmitters** (Key 47) released by neighboring neurons turn off the sodium pump, resulting in a **depolarization** of the membrane. This is more than enough to interrupt the sodium pump in the next section of membrane, which then depolarizes, and the

process repeats itself rapidly along the membrane, resulting in a **nerve impulse**, which is essentially a **wave of depolarization**.

- Electrical recordings show a reverse polarization of characteristic shape called a **spike** or **action potential**.
- After the nerve impulse has passed, the sodium pump resumes its action and reestablishes normal polarity after a short delay.

Neuroglia: Cells of the nervous system, other than neurons. These cells provide structural support and nutrition to the neurons. Most nervous tissue is avascular (blood vessels lie on the outside only), so nutrients must diffuse in and wastes must diffuse out. The neuroglia aid in this process, often by sending out a **perineural foot** to surround the neuron and a **perivascular foot** to surround a nearby blood vessel. *Examples*: Astrocytes, oligodendrocytes (oligodendroglia), microglia, Schwann cells.

Organization of the nervous system:
- **Central nervous system (CNS):** Brain and spinal cord.
- **Perpheral nervous system (PNS):** Peripheral nerves, including both **cranial nerves** (originating in the brain) and **spinal nerves** (originating in the spinal cord).
- **Special sense organs:** Eye, ear, taste buds, nasal epithelium.
- **Nerves** are bundles of axons outside the CNS; **tracts** are similar bundles within the brain.
- **Ganglia** are clumps of cell bodies outside the CNS; **nuclei** are similar clumps in the brain.

Key 47 Neurotransmitters

OVERVIEW *Neurons release **neurotransmitters** where they meet other neurons (at **synapses**) or where they meet muscles. Possible effects, either stimulation or inhibition, occur by changing membrane characteristics.*

Neurotransmitters: Neurons release chemicals called **neurotransmitters** where they meet other neurons (at **synapses**) or where they meet muscles. Some neurotransmitters **stimulate** postsynaptic cells by depolarizing membranes; others **inhibit** cells by increasing membrane polarity. Most neurotransmitters accumulate in membrane-bounded **vesicles** in the presynaptic neuron. They are released by **exocytosis** (Key 12). Some neurotransmitters (dopamine, serotonin) are recycled: the cell that secreted them resorbs them (**re-uptake**) and reuses them.

Acetylcholine (AC or ACh): The most common neurotransmitter, secreted along with an enzyme (**cholinesterase**) that degrades it. It occurs at all synapses in the peripheral nervous system and is stimulatory; it also stimulates muscle contraction and slows the heartbeat.

Norepinephrine and epinephrine: Released in the brain and by certain nerves (**sympathetic nerves**). They are stimulatory, quickening the heartbeat and increasing blood supply to most muscles.

Dopamine: Generally stimulatory; released primarily in the brain.

Serotonin (5-hydroxytryptamine, 5HT): Stimulatory, formed only in the brain. More of it is made, released, and recycled during dreams.

Gamma-amino butyric acid (GABA): Inhibitory.

Glutamate and glycine: Amino acids which may also be neurotransmitters.

Endorphins and enkephalins: Natural painkillers in the brain; may also function as neurotransmitters.

Key 48   Central nervous system

OVERVIEW   *Most invertebrates have nervous systems derived from the ladder-like arrangement in flatworms. Vertebrate brains develop in three portions (forebrain, midbrain, hindbrain). In mammals, the cerebral hemispheres enlarge, and their size and complexity become a crude measure of intelligence.*

Invertebrate nervous systems: Cnidaria have a **nerve net** of interconnected neurons with no center. Flatworms have two long chains of ganglia in a ladder-like arrangement; the largest ganglia, near the eyes, form the beginnings of a ''brain.'' Most other invertebrates use modifications of this ladder-like pattern; a major nerve cord runs along the ventral midline, splits to form an **esophageal ring**, and reunites above the mouth to form a **cerebral ganglion** or **brain**.

Embryonic vertebrate brains: Form as three major divisions:
* **Forebrain (prosencephalon)**, primitively devoted to smell.
* **Midbrain (mesencephalon)**, primitively concerned with vision.
* **Hindbrain**, dealing with sound and vibrations.

Adult vertebrate brains: Organized into five regions:
* **Telencephalon:** Paired parts of the forebrain, including **olfactory bulb**, **olfactory lobe**, and **cerebral hemispheres**, which enlarge greatly in mammals and take over many added functions.
* **Diencephalon:** Unpaired, second portion of the forebrain, including the **pineal body** or **epiphysis**; **tela choroidea** (thin roof); **thalamus** (controls many emotions); **hypothalamus** (controls appetite and body temperature); and part of **pituitary gland** (Key 52).
* **Mesencephalon:** Midbrain, including **corpora quadrigemini**.
* **Metencephalon:** Includes **cerebellum** and **pons**.
* **Myelencephalon:** **Medulla**, continuing into the spinal cord.

Brain ventricles: Cavities containing **cerebrospinal fluid**.

Spinal cord: **White matter** (myelinated **tracts**) and **gray matter** (unmyelinated motor and sensory columns).

Spinal reflex pathway: **Sensory neuron** runs from **receptor cell** in skin to cell body in **dorsal root ganglion**, then into somatic motor column of spinal cord. **Association neuron** connects somatic motor to somatic sensory column. **Motor neuron** runs from somatic motor column out ventral root to a voluntary muscle or other **effector cell**.

Key 49   Sense organs

OVERVIEW *Nerve endings in the skin detect five different sensations: touch, pressure, cold, warmth, and pain, each sensed by a different type of nerve ending. There are many types of eyes; vertebrates have eyes in which a lens focuses light upon a sensitive retina. Lateral lines in fishes detect underwater swimming movements. The outer ear focuses sound onto the tympanic membrane. The middle ear has one to three vibrating bones (ossicles). The inner ear is filled with fluid and aids in both hearing and balance.*

Interoceptors: Nerves that detect the status of the body's internal organs.

Proprioceptors: Detect the position of the body's muscles and joints.
- **Neuromuscular spindles** detect the state of contraction of muscles.
- **Stretch receptors** detect tension in ligaments or tendons.
- **Position receptors** (in arthropods) detect angular position of joints.

Exteroceptors: All other receptors, responsive to external conditions:
- **Thermoreceptors** are sensitive to temperature.
- **Mechanoreceptors** detect mechanical stimuli, like vibrations and sound.
- **Photoreceptors** are sensitive to light.
- **Chemoreceptors** are sensitive to chemical stimuli (taste, smell).

Cutaneous sensations: Nerve endings in skin detect five different sensations:
- **Naked (unencapsulated) nerve endings** perceive pain.
- **Meissner's corpuscles** are sensitive to light touch.
- **Corpuscles of Pacini** are sensitive to deep pressure.
- **Ruffini corpuscles** are sensitive to warmth.
- **Krause end bulbs** are sensitive to cold.

Smell: The **nasal epithelium (olfactory mucosa)** is chemically sensitive to particles in very low concentrations. The sensory cells are unusual because they grow inward toward the olfactory lobe of the brain.

Taste: The **taste buds** are sensitive to moderate concentrations of chemicals.

Photoreception (vision): Many simple organisms detect light and dark without any special organs. Some protists, like *Euglena*, have photo-sensitive "pigment spots." Flatworms and many other simple animals have light-sensitive optic nerves, often with an overlying transparent skin.

Image-forming eyes use various optical principles:
- "Compound eyes" (as in insects) use an array of brightness-and-color receptors to form an image, like a TV picture, out of many dots.
- Pinhole eyes focus light through a very small hole onto a sensitive retina.
- Lens-based eyes (as in molluscs and vertebrates) use a transparent lens to focus an image upon a retina.

Vertebrate eyes include:
- **Eyelids:** Protective folds of skin.
- **Cornea:** Transparent layer protecting front of eye.
- **Lens:** Focuses light upon retina.
- **Ciliary body:** Holds lens in place and controls focus of image.
- **Iris diaphragm:** Controls amount of light entering eye.
- **Aqueous humor:** Watery fluid in front of lens.
- **Vitreous humor:** Glassy fluid behind lens.
- **Retina:** Light-sensitive surface, containing rods and cones.
- **Choroid and scleroid coats:** Protective layers around the eye.
- **Optic nerve:** Carries impulses to the brain.

Stato-acoustic senses: Special forms of mechanoreception.
- **Lateral line** in fishes: Sensitive to swimming movements of other fishes or to waves reflected from obstacles.
- **Outer ear** (in mammals): Focuses sound waves on tympanic membrane.
- **Middle ear:** Contains tiny bones (ossicles) that transmit vibrations. Mammals have three ossicles: hammer (malleus), anvil (incus), and stirrup (stapes); other land vertebrates have stapes only.
- **Inner ear: Cochlear portion** contains **cochlea** (coiled in mammals), including vibration-sensitive **organ of Corti. Vestibular portion**, sensitive to balance and acceleration, contains semicircular canals.

Key 50   Hormones in general

OVERVIEW *Endocrine glands secrete hormones directly into the blood. Endocrine organs act more slowly than the nerves. Specific hormones are listed in Keys 51 and 52. Most hormones activate target cells by a common mechanism which uses adenyl cyclase to make cyclic AMP.*

Endocrine secretion: Secretion of substances directly into the bloodstream. Anything secreted this way is called a **hormone**. Hormones must have an effect on a target tissue. They are carried by the blood, so their targets can be in any location or in many places.

Response over time: Endocrine glands usually act much slower than the nervous system. Hormone levels may build up over time to control slow changes such as sexual development, growth, or metamorphosis.

Embryological sources: Since the bloodstream distributes hormones, endocrine glands can be located anywhere and derived from any source:
- From endoderm: Thyroid, parathyroids, intestinal lining, pancreas.
- From mesoderm: Adrenal cortex, ovary, testis.
- From ectoderm: Anterior and posterior pituitary, adrenal medulla.

Hormone activation: Most hormones activate their target cells by a common mechanism, as follows:
1. The hormone binds to a specific receptor on the target cell surface. Hormone specificity depends on which target cells have receptors for which hormones and on how these target cells respond.
2. The receptor activates the enzyme adenyl cyclase (=adenylate cyclase), which converts ATP into **cyclic AMP (cAMP)**.
3. The cyclic AMP acts as a "second messenger," carrying information from the cell membrane to the nucleus. The nucleus then responds by turning certain genes "on" or "off" (Key 35).

Plant growth substances (auxins, giberellin, Shikimic acid, etc.): Plant growth substances bear many similarities to animal hormones. Auxins (like 5-hydroxy indole acetic acid, HIAA) cause plant stems to elongate. In stems lighted from the side, auxins cause the dark side to elongate, so the plant turns its top toward the light source.

Key 51   Steroid hormones

OVERVIEW   *Steroids are substances derived from cholesterol. Steroid hormones control sexual maturation, sexual cycles, mating seasons, and insect metamorphosis.*

Primary sexual characteristics: Those which are absolutely necessary for reproduction, like a uterus or egg production.

Secondary sexual characteristics: Those which may help attract a mate but which are not absolutely necessary for reproduction to occur, like deep voices, beards, antlers, or peacock feathers.

Seasonal variation: In animals that mate seasonally, the seasonal production of sex hormones controls the maturation and regression of sex characteristics and sexual behavior.

Androgenic (masculinizing) hormones: Testosterone, secreted by testes, causes development of primary and secondary male characteristics. Androgens produced by the adrenal cortex cause development of secondary male characteristics only.

Estrogen: Female hormone, secreted by ovaries, which causes development of primary and secondary female characteristics.

Progesterone: Female hormone secreted by ovaries after ovulation.

Adrenal cortex hormones: Outermost layer produces **glucocorticoids**, affecting some aspects of sugar metabolism. Middle layer produces **mineralocorticoids** such as aldosterone, controlling metabolism of such ions as Na^+. Inner layer produces androgenic hormones.

Insect hormones:
- **Ecdysone** (molting hormone, a steroid) controls molting and insect metamorphosis.
- **Juvenile hormone** controls the outcome of molting: if present in a larva, the molt produces a bigger larva, but absence of juvenile hormone causes molt to result in metamorphosis into an adult.
- **9-keto-decanoic acid:** secreted by queen bees and fed to workers and larvae to keep them sterile. If queen dies, absence of this substance causes some new larvae to develop into a new queen.

Key 52 Peptide hormones

OVERVIEW *Many hormones are short peptide sequences or small proteins. The pituitary is sometimes called the "master gland" because it produces so many hormones, including those that control other endocrine glands. The thyroid and parathyroid glands produce hormones that control metabolic rate and calcium metabolism. The adrenal medulla produces hormones that stimulate a "fight or flee" response. The pancreas secretes hormones controlling glucose metabolism, and other parts of the digestive tract produce hormones to regulate digestive functions.*

Anterior pituitary gland: Derived from tissue that detaches from the roof of the mouth and lodges at the base of the brain. Many of the hormones produced by this gland are called **trophic hormones** because they stimulate other glands to release other hormones.
- **Follicle-Stimulating Hormone (FSH):** Stimulates growth and maturation of the gonads, including secretion of their hormones.
- **Luteinizing Hormone (LH):** Induces ovulation in females.
- **Luteotrophic Hormone (LTH):** Stimulates progesterone production.
- **Adrenal CorticoTrophic Hormone (ACTH):** Stimulates adrenal cortex to produce its hormones.
- **ThyroTrophic Hormone:** Stimulates the production of thyroxin.
- **Growth hormone (somatostatin):** Stimulates growth.
- **Melanocyte Stimulating Hormone (MSH):** Stimulates pigment cells.

Posterior pituitary hormones:
- **Vasopressin (anti-diuretic hormone, ADH):** Stimulates urinary retention of water and concentration of urine; lack of hormone causes excessive levels of urination, thirst, and drinking.
- **Oxytocin:** Stimulates uterine contractions in childbirth.

Pineal gland: Secrets **melatonin** during darkness, which regulates circadian rhythms (Key 53) and clues the body to day length and seasonal changes.

Adrenal medullary hormones: The medulla of the adrenal gland secretes **epinephrine** ("adrenalin") and **norepinephrine**. These hormones are also neurotransmitters (Key 47). They stimulate a more rapid heartbeat and breathing, more blood flow to muscles (and less to digestive tract), lowering of nerve threshholds (so nerves fire more easily), and shortening of blood clotting time. Overall, they prepare the body to "fight or flee."

Thyroid and parathyroid hormones:
- **Thyroxin:** An iodine-containing protein produced by the thyroid gland. It increases metabolic rate.
- **Calcitonin** (produced by thyroid gland): Stimulates bone deposition and depletion of calcium from blood.
- **Parathyroid hormone:** Stimulates release of calcium from bones and increase of blood calcium levels.

Gastrointestinal hormones:
- **Gastrin:** Stimulates gastric secretion.
- **Secretin:** Stimulates pancreatic and liver secretion.
- **Pancreozymin:** Stimulates more concentrated pancreatic secretions.
- **Enterogasterone:** Delays emptying of stomach contents.
- **Cholecystikinin:** Stimulates gall bladder to release bile into intestines.

Pancreatic hormones (produced by **islets of Langerhans**):
- **Insulin:** Aids in glucose metabolism and glycogen breakdown; deficiency produces **diabetes mellitus**.
- **Glucagon:** Opposite in effect to insulin, stimulates glycogen storage and depletion of glucose from blood.

Key 53   Biological rhythms

OVERVIEW   *Biological clocks of about 24 hours' duration are called **circadian clocks**. Many body functions follow circadian rhythms controlled by light intensity. Changes in day length can control seasonal responses (such as flowering, mating, or migration) in both plants and animals.*

Circadian clocks: Many organisms maintain "biological clocks" that control rhythms approximately 24 hours in duration (**circadian rhythms**). In vertebrates, the **pineal body (epiphysis)** appears to control such rhythms in response to changes in light intensities of many hours' duration. Body temperature, hormone levels, neurotransmitter levels, and immune responses all follow circadian rhythms. In constantly dim light, these rhythms continue for days, though they slowly drift away from exactly 24-hour duration. Bright lights reset circadian clocks to conform to natural day/night rhythms. Rapid resetting of the clock by more than about an hour produces "jet lag," increasing the need for rest and depressing the immune system.

Seasonal responses (photoperiodism) in plants: Plants need to flower or their seeds need to sprout at particular times of the year. Pollen production must occur at the same time in all members of a population if pollination is to occur. The timing of these events often depends on increasing or decreasing day length and can be artificially manipulated by gradually changing circadian rhythms of light and darkness. "Long day" (or "short night") plants flower in response to dark periods below a certain critical duration. "Short day" (or "long night") plants require dark periods above a certain duration before flowering; brief flashes of brightness will interrupt these dark periods and inhibit flowering.

Seasonal responses (photoperiodism) in animals: Mating seasons, including seasonal changes in hormone levels and secondary sexual characteristics, are controlled in many species by changing day lengths. Artificial increases in the duration of daytime light intensities produce springtime responses; artificial decreases in daylight produce fall-like responses. Seasonal migrations in migratory species are also controlled this way, as is fat deposition in species that need to store up fat for migration or for overwintering. Artificial shortening or lengthening of dark periods can induce migration in seasonally inappropriate directions, accumulation of winter fat, and similar responses.

Key 54 Animal behavior

OVERVIEW *Simple behaviors include growth movements (tropisms) and such locomotor movements as taxes and kineses. Complex innate behaviors are called instincts; they tend to be inflexible, and no time is wasted in learning them. Learned behaviors vary more with circumstances, but a learning period is a necessary prelude. Symbolic behavior and language are keys to human-like intelligence.*

Tropisms: Growth or turning movements in plants and sessile animals. Tropisms and taxes directed *toward* a stimulus are called **positive**; those directed *away from* the stimulus are **negative**.
- **Phototropism:** Growth or turning toward light (positive phototropism) or away from light (negative phototropism).
- **Geotropism:** Growth toward or away from the earth's center.
- **Chemotropism:** Growth toward or away from a chemical; special types include **halotropism** for salt and **hydrotropism** for water.
- **Anemotropism:** Growth toward or away from a source of wind.
- **Thigmotropism:** Growth toward or away from something touched.

Taxes: Oriented locomotor behaviors in motile organisms.
- **Phototaxis:** Locomotion toward or away from light.
- **Geotaxis:** Locomotion toward or away from the earth's center.
- **Chemotaxis:** Locomotion toward or away from a chemical; special types include **halotaxis** for salt or **hydrotaxis** for water.
- **Anemotaxis:** Locomotion upwind (positive) or downwind (negative).
- **Rheotaxis:** Swimming upstream (positive) or downstream (negative).

Kineses: Non-oriented locomotion, with no particular direction.
- **Photokinesis:** Locomotion in response to light.
- **Chemokinesis:** Locomotion in response to chemicals.
- **Thigmokinesis:** Locomotion in response to touch.

Instincts: Complex, innate behavior patterns. ''Complex'' means that several acts need to be done in proper succession. ''Innate'' means inborn or genetically programmed. A standard test is whether animals raised alone from birth can perform the behavior correctly. Instincts are usually **stereotyped**, meaning that the behavior does not vary from one occasion to another or from one performer to the next (good for courtship and species-recognition behavior).

- Advantage: No time or effort is wasted in learning or in making mistakes; behavior is correct the first time.
- Disadvantages: Cannot be modified to suit circumstances. *Examples*: Most aggressive or submissive postures and movements; courtship and mate-attracting behaviors (including songs, etc.); nest-building behavior in many species; web-weaving in spiders; territoriality.

Learned behavior: Behavior that improves with practice. Learned behavior is more variable (an advantage in interactions with the local environment, but a disadvantage in mate-recognition or courtship behavior).

- Advantages: Can be varied to suit local circumstances; can become more complex than instincts.
- Disadvantage: Learning (and mistakes) must take place first. The amount of possible learning is limited by neural complexity.

Imprinted behavior: Behavior learned very early in life.

Conditioned learning: Any behavior with a pleasant result will be reinforced and repeated; behavior with an unpleasant result will tend not to be repeated. If behavior has different outcomes depending on pre-existing stimuli, subject will learn to discriminate on the basis of those stimuli.

"Insight" and "rational" learning (principally in mammals): Solution occurs all at once, in "flash of insight" (also called "aha!"). Much learning occurs in play and exploratory behavior and by imitation.

Symbolism and language: Primates can be taught that a stimulus has an arbitrary meaning (it stands for something else). The highest development of symbolic behavior is language.

Learning and intelligence: Intelligence is a capacity for learning increasingly complex behaviors. It correlates with brain size only if species of comparable size are compared. Mammals are generally the most intelligent animals.

Theme 8   ECOLOGY

*E*cology is the study of species in relation to their environment. Ecologists study the growth of populations and the interactions among species through mathematical models. Temperature, rainfall, wave action, and other physical charateristics can determine what species can live in certain environments, what adaptations they must possess, and what species cannot survive. Groups of species that occur together interact to form communities; together with the physical environment they form ecosystems. Species also have characteristic geographic ranges that can be grouped into biomes and regions; biogeographers study the ways in which these distributions can change with time.

Key 55   Ecology: Populations

OVERVIEW   *Populations include all those members of a species who can interbreed. The role of a population in the ecosystem is called its niche; the place where it lives is called its habitat. The simplest model of population growth is exponential growth without limit. A more realistic model assumes that population size reaches an upper limit (carrying capacity) in each environment. Populations that are r-selected reproduce prolifically but offer no parental care. Populations with stable population sizes near the carrying capacity are K-selected and provide parental care.*

Population: All members of a species who can interbreed or exchange genes with one another and leave fertile offspring.

Habitat: The place where a population lives.

Niche: The role or ''occupation'' or ''way of life'' of a population; its position in the workings of the ecosystem.

Exponential growth: Assume a population increases by a proportion r in each generation. Then, using calculus notation,

$$dN/dT = rN$$

where N = population size and r is a constant called the Malthusian parameter or ''intrinsic rate of natural increase.'' This equation can be solved as follows:

$$N = N_O \, e^{rT}$$

where N = current population size, N_O = initial population size, e = base of natural logarithms, r = intrinsic rate of natural increase, and T = time (in generations). Under this model, population size keeps increasing without limit.

Logistic growth: A more realistic model assumes that population size eventually reaches an environmentally determined limit, the ''carrying capacity,'' after which mortality limits further growth. In this model,

$$dN/dt = r \, N \, (K-N)/K$$

where K is the carrying capacity and other symbols are defined already. The resulting S-shaped (sigmoid) growth curve levels off as population size (N) approaches K.

Density-dependent controls: Those whose effects increase with population size. *Examples:* limitations on food and space, increased stress and aggression at higher densities, more predation and infectious diseases.

Density-independent controls: Those that act independently of population size. *Examples:* mortality by floods, earthquakes, weather extremes, or landslides and other habitat destruction.

r-selection: Populations suffering frequent, devastating density-independent mortality are **r-selected**. Most r-selected species have population sizes well below carrying capacity, selection favoring rapid population growth (large *r*), prolific reproduction, little or no parental care, heavy mortality to newly released eggs or other initial stages, short individual life span, and early reproduction at small sizes.

K-selection: Populations suffering little or no density-independent mortality (like humans and many other mammals) are **K-selected**. Most K-selected species have populations at or near carrying capacity (K), selection favoring economy of resources and efficiency (''making do'' with less food or less space), long individual life span, and reproduction only at larger sizes. Fewer eggs or offspring are produced, but more stored food or parental care is given to each. Most mortality is density-dependent (from crowding, etc.) and occurs throughout life, especially at older ages.

Key 56 Ecology: Species interactions

OVERVIEW *Types of interaction include competition, in which each species inhibits the population growth of the other. In predation and parasitism, one species (the predator or parasite) benefits from interacting, while the other is harmed (its population size diminishes). Many cases of parasitism evolve toward commensalism (in which the host population is neither harmed nor benefitted) and eventually mutualism (in which both species benefit). Other frequent results of interactions among species include niche subdivision and specialization.*

KEY TYPES OF INTERACTIONS

In the following chart,
+ means a benefit (or population increase) from interaction
− means a harm (or population decrease) from interaction
0 means that the interaction has no effect on population size

	Effect on species 1	Effect on species 2	
Competition	−	−	Inhibits growth of both species
Amensalism	−	0	Only one species inhibited
Neutralism	0	0	No effect on either species
Predation & parasitism	+	−	One species benefits, the other is harmed
Commensalism	+	0	Benefits one species, neutral to the other
Mutualism	+	+	Benefits both species

Competition: Occurs when each species reaches a smaller population size than it would in the absence of the other. Most often, both species are competing for a limited resource like food or space. Competition is described mathematically by the equations of Lotka and Volterra:

For species 1: $dN_1/dT = r\,N_1\,(K_1 - N_1 - aN_2)/K_1$
For species 2: $dN_2/dT = r\,N_2\,(K_2 - N_2 - bN_1)/K_2$

where N_1 is the population size and K_1 the carrying capacity for species 1 and N_2 and K_2 are the analogous quantities for species 2.

Depending upon conditions, there are four possible results:
- Species 1 always wins (species 2 becomes locally extinct)
- Species 2 always wins (species 1 becomes locally extinct)
- Either species can win, depending upon initial population sizes: whichever gains the upper hand wins; the other becomes extinct.
- Both species will coexist indefinitely.

Exploitation (including both **predation** and **parasitism**): One species (the **prey** or **host**) has a beneficial effect on the other (the **predator** or **parasite**, whose population increases), but the second species causes a population decrease in the first. Parasites tend to be much smaller than their hosts, but predators are similar in size to their prey or larger. Many simple predator-prey situations (one predator species, one prey) are unstable: if the predators kill many prey, they will then starve (unless they can switch to another food supply). Some predator-prey interactions result in fluctuating cycles of abundance, but most are stabilized by the tendency of predators to switch from one prey species to another, choosing whichever prey is most abundant.

Symbiotic interactions: Species living together in close contact, often one in or upon the other, which serves as its **host**.

Parasitism: The parasite benefits from the interaction, while the host is harmed. Most parasites are r-selected. Many parasites have hooks, suckers, or other adaptations for holding on. Highly evolved parasites are very small or very flat and often have complex life cycles. Flat shapes facilitate absorptive nutrition and hinder dislodging. Small size and inconspicuous entrance are common adaptations to avoid detection and minimize harm to host. Organs that are dispensable, especially sense organs, nervous system, and most organs of nutrition, respiration, and excretion, are lost in highly evolved parasites.

Commensalism: One species benefits; the host receives equal benefit and harm, so its population size neither increases nor decreases.

Mutualism: Both species benefit from the interaction.

Results of species interaction:
- Species may specialize, subdividing large niches into smaller ones, each exploited by a different species.
- The interaction may evolve to a more favorable form of interaction, e.g., from parasitism to commensalism to mutualism.
- Species that are less successful may find new ways of life (new niches) in which they might succeed.
- Species that can do none of these things become extinct.

Key 57   Ecology: Communities

OVERVIEW   *A community consists of all the species that live together and interact in a particular habitat. Energy moves through food chains from primary producers to primary consumers to secondary and higher-order consumers, forming a food pyramid with four or fewer trophic levels.*

Community: All the species that live together and interact in a particular habitat, like a particular bog or a particular coral reef.

Biomass: The total quantity of living matter in a community.

Food chains: Each community has **primary producers**, or **autotrophs**, who can derive energy from sunlight (or chemicals) and carbon compounds from atmospheric CO_2. Most autotrophs are photosynthetic plants; their rate of synthesis of new biomass is called **primary production. Heterotrophs** include all species that need to feed on other organisms.
- Primary producers are eaten by **primary consumers (herbivores)**.
- Primary consumers are eaten by **secondary consumers (carnivores)**.
- Secondary consumers are eaten by **higher order consumers**.

Food webs: Many food chains are so intricate that they form **food webs**.

Food pyramid: Each **trophic level** has less biomass and less total energy than the level that supports it. Ratios between levels reflect the low efficiency of energy conversion. The great energy loss at each new level limits the maximum number of levels to about four.

Biomagnification: A chemical that is not broken down (like a pesticide or toxic metal) reaches higher concentrations as it ascends the food pyramid. This is because the amount of the chemical, largely unchanged, is distributed in a smaller biomass at each level.

Succession and climax: When a new habitat appears (e.g., after a forest fire or the birth of a new island), the first colonizers form a community, which makes new niches available for new species to come in. Some of the newcomers may force earlier species out of existence. In this process, called community **succession**, communities keep replacing one another until enough species resist displacement and form a **climax community**.

Key 58   Ecology: Ecosystems

OVERVIEW   *An **ecosystem** includes both a community and the physical environment surrounding and interacting with it. The largest of all ecosystems is the **biosphere**— planet Earth and all its inhabitants.*

Physical environments: Environments differ in temperature, humidity, rainfall, and seasonal variation in all these factors; also elevation, slope, and soil type, salinity of water, and water chemistry.

Adaptations to extreme environments: Species that adapt to extreme environments generally enjoy freedom from competition with other species that cannot cope with such environmental stresses.
- Adaptations to cold include insulation, leaf loss, and less surface.
- Adaptations to extreme heat and dryness (desert conditions) may include burrowing, daytime inactivity, or thick, waxy cuticles.

Carbon cycle: Autotrophs (mostly green plants) fix CO_2 into organic compounds, which then pass up the food chain. Decomposer organisms (**saprobes** or **saprophytes**) feed on dead or decaying organic matter (corpses, feces, fallen leaves, etc.). All organisms return CO_2 to the air and other carbon compounds to the soil and water.

Oxygen cycle: All organisms (except for anaerobic bacteria) use oxygen in their respiration and release CO_2 and water. Photosynthetic plants split water molecules and release oxygen to the atmosphere.

Nitrogen cycle: Nitrogen-fixing bacteria convert atmospheric N_2 into soluble nitrates, which plants absorb. Nitrogen compounds (proteins, etc.) pass up the food chain. Animals excrete urea and other nitrogen wastes. Saprobes digest proteins into simpler compounds. **Nitrifying bacteria** convert ammonium salts into nitrites, then into nitrates. **Denitrifying bacteria** convert nitrates to N_2 gas.

Ecological problems:
- **Pollution** results from problems of waste disposal.
- **Acid rain** results from burning sulfur (an impurity in coal).
- **Destruction of tropical forests** and **burning of fossil fuels** produce a global CO_2 increase, causing a **greenhouse effect**, which results in **global warming. Ozone depletion** from certain photoactive chemicals reaching the upper atmosphere also hastens global warming.
- **Endangered species** risk extinction from habitat destruction.
- **Overpopulation** can worsen all of the above problems.

Key 59   Biogeography

OVERVIEW   *Animal and plant distributions follow many global patterns. Ecologically similar species form similar communities (biomes). Another pattern is for families to be confined to certain land masses, forming regions separated by barriers to dispersal. Animal and plant distributions thus reflect both adaptations and dispersal capabilities. Large islands have more species than small ones.*

Animal and plant distributions: All the animals of a habitat or region consitute its **fauna**; all the plants constitute its **flora**.

Biomes: Ecologically similar communities with ecologically similar species showing convergent adaptations. Biomes include **tundra** (treeless and cold), **taiga** (northern conifer forests), **temperate deciduous forest** (trees shed leaves annually), **temperate grassland** (grain-rich prairies), **chaparral or Mediterranean** (mild, rainy winters), **desert**, **tropical rainforest** (many tall trees, great variety of species), **tropical deciduous forest**, **tropical scrub forest** (scant rainfall, many thorny bushes), **tropical grassland (savanna)**, **freshwater** streams and lakes, and **marine** biomes (intertidal, etc.), arranged by depth.

Regions: Continental areas inhabited by related species (belonging to the same family), although ecological adaptations may differ. The six zoogeographic regions include the **Nearctic** (Alaska to central Mexico), **Neotropical** (the Americas from Mexico to Cape Horn), **Palaearctic** (North Africa and most of Eurasia), **Ethiopian** (Africa south of the Sahara), **Oriental** (tropical Asia), and **Australian** (Australia, New Zealand, and Pacific islands).

Dispersal and colonization: Most species can disperse freely across continents, limited only by the need to adapt to new environments. **Land bridges** (like Panama) allow dispersal in both directions. **Sweepstakes dispersal** is the occasional crossing of barriers by high-risk means, usually in one direction only by only a few species.

Island biogeography: Many forces that control animal and plant distributions are easier to study on islands. Islands receive most of their species from the nearest continent. Larger islands are more frequently colonized by new species, but they also suffer more competition, and thus more local extinction of species. A frequent pattern, the **species-area curve**, $S = c A^z$, relates species diversity to island size.

Theme 9 EVOLUTION

*E*volution is the process of long-term change in biological systems. The simplest form of evolution consists of changes in gene frequencies within populations brought about by mutation, natural selection, genetic drift, and other processes. These processes produce changes within populations and differences among populations. New species arise from the splitting of older species, a process which requires reproductive isolation. Paleontologists study long-term changes by examining the historical record of life preserved as fossils. These changes are responsible for several types of resemblance among organisms, for adaptive radiation, for evolutionary trends, and for the pattern of relationships that results.

Key 60   Evolution: Overall theories

OVERVIEW   *Until 1859, most biologists believed in the fixity of species. Lamarck and Geoffroy thought that species could adapt to their environments by changes occurring within individual lifetimes. Darwin proposed instead that natural selection resulted in "descent with modification."*

Historical ideas before Darwin: Most scientists and philosophers before 1859 insisted that species were fixed and unchanging; each was thought to be a copy of an unchanging heavenly form (=type, idea, *eidos*). Species were arranged in an unchanging, unbroken order of perfection, the **great chain of being**.

- **J.B. Lamarck** tried to explain environmental adaptations. He claimed that the willful use of a part would strengthen and enlarge it, while the disuse of a part would cause it to wither (**Lamarckism**, or the theory of use and disuse). Lamarck also believed in a single-file line of progress ("la marche de la Nature").
- **E. Geoffroy Saint-Hilaire** explained adaptation by direct effects of the environment (**Geoffroyism**), a concept which Lamarck rejected.
- Both Lamarck's and Geoffroy's theories rely upon the inheritance of **acquired characteristics**, changes occurring during an organism's lifetime. This possibility was later disproved by Weismann and others.

Charles Darwin (1809–1882):   A voyage around the world convinced Darwin of several facts that earlier theories could not explain:

- Different continents had very different species, even in similar climates.
- Species sharing a land mass or island group were often related.
- Island species were usually related to those of the nearest continent.
- Similar environments did not always produce the same species or related species (contrary to Lamarckism and Geoffroyism).

***On The Origin of Species* (1859):**   In this book, Darwin explained all of the above facts by **"descent with modification,"** a branching form of evolution very different from earlier theories. This book also suggested **natural selection** as a mechanism for evolutionary change.

- All living species tend to over-reproduce.
- Most seeds, eggs, or hatchlings die without reproducing.
- All living species are extremely variable.
- Many of these variations are inherited.

- Inherited differences in survival and reproductive ability (**natural selection**) bring about change in each generation.

Evidences for evolution ("descent with modification"):
- Patterns of common descent are reflected in classifications.
- Related species share many internal similarities (anatomical, biochemical, or embryological **homologies**) despite different adaptations.
- These homologies may include **vestigial** remnants of once-useful parts.
- Similar adaptations often occur under similar circumstances, even in unrelated species (**convergent adaptations**).
- Related species often inhabit certain land masses or island groups.
- Fossils can often be arranged in evolutionary sequences.
- Some species vary from place to place, and the differences are inherited.

Evidences for natural selection:
- All living species are highly adapted to their ways of life.
- Many adaptations cannot be explained by environmental influence alone. *Examples:* Unrelated but ecologically equivalent species live on different continents; some embryonic structures (e.g., a flap in the fetal human heart that seals closed at birth) develop before they become useful; some behavior (like bird migration or nest building) occurs in advance of its usefulness.
- Some adaptations are less than perfect, contrary to an earlier theory that used perfect adaptation to prove divine creation.
- Natural selection has sometimes been documented (e.g., among peppered moths in England), and has resulted in changes in natural populations.
- Artificial selection by animal and plant breeders has produced many new adaptations, some of them similar to adaptations occurring naturally.

Mimicry and camouflage: Many species gain protection against predators by resembling their background (**camouflage**) or by falsely resembling other species (**mimicry**). In **Batesian mimicry**, a palatable species resembles a distasteful or harmful one. **Müllerian mimicry** is resemblance among distasteful or harmful species.
- Mimicry works only when certain **models** are present, a fact explained easily by natural selection, but not by Lamarckism or similar theories.
- Mimicry may vary geographically, with the same mimic species resembling different models in different localities. Natural selection can explain this; Lamarckism cannot.

Key 61   Genes in populations

OVERVIEW   *Large, random-mating populations will, under certain assumptions, reach a genetic equilibrium in which genotypic proportions tend to remain constant.*

Hardy-Weinberg law: In a large, random-mating population of diploids with no unbalanced mutation, unbalanced migration, or selection in any form, the genotypic proportions tend to remain constant. This constancy is called a **genetic equilibrium**, with equilibrium frequencies given by the equation

$$p^2 AA + 2pq\ Aa + q^2\ aa = 1$$

- p stands for the frequency of *A*; q for the frequency of *a*; p + q = 1.
- p^2 is the frequency of *AA* homozygotes; gametes are all *A*.
- 2pq represents the frequency of heterozygotes (*Aa*). Half of their gametes (pq) are *A*; the other half (pq) are *a*.
- q^2 is the frequency of *aa* homozygotes; gametes are all *a*.
- To find the new frequency of allele *A*, add $p^2 + pq = p(p + q) = p$; so the frequency of allele *A* remains p.
- To find the new frequency of allele *a*, add $pq + q^2 = (p + q)q = q$; so the frequency of allele *a* remains q.
- A Hardy-Weinberg equilibrium can be established in a single generation of random mating.

Exceptions to the Hardy-Weinberg law:
- If the population is not large, **genetic drift** occurs: gene frequencies can fluctuate randomly in either direction simply by chance.
- Populations may not mate at random. **Inbreeding** (increased mating among related individuals) results in more homozygotes. **Assortative mating** is mating according to phenotype, with mating between phenotypically similar individuals being either more frequent (**positive assortment**) or less frequent (**negative assortment**).
- Mutation in one direction only can cause one allele slowly to replace another. Mutation in both directions results in an equilibrium with frequencies determined by the mutation rates.
- Migration between populations always causes the gene frequenies of the receiving population to shift toward those of the immigrants.
- **Selection** (Key 62) occurs whenever the several genotypes contribute genes unequally to the next generation.

Key 62   Selection and microevolution

OVERVIEW   *Evolution below the species level results from variation and from the action of natural selection and other forces on this variation. Selection occurs when chances of leaving offspring differ among genotypes.*

Microevolution: Evolution below the species level. It results from:
- **Variation**, brought about by mutations, chromosomal changes (Key 30), and by genetic recombination through mating.
- Selection, genetic drift, and other forces that act upon variation.
- Reproductive isolation (Key 63).

Selection: Genotypes contribute genes unequally to the next generation.
- **Natural selection** is selection by natural processes. The peppered moths of England, selected by birds, are an example.
- **Artificial selection** is selection of captive species by humans.
- **Sexual selection** is selection based on success in mating.
- Selection against a dominant trait can eliminate the trait rapidly.
- Selection against a recessive trait works very slowly and becomes much less effective once the recessive gene becomes rare.
- Selection against heterozygotes can result in either allele becoming lost and the other taking over 100% of the gene pool.
- Selection favoring heterozygotes over both types of homozygotes results in **balanced polymorphism** in which both alleles persist indefinitely. Sickle-cell anemia is an example of this situation.
- **Directional selection** shifts the population mean.
- **Disruptive selection** increases population variance.
- **Centripetal or stabilizing selection** reduces variance.

Geographic variation: Natural selection in different environments causes populations to differ. **Gene flow** reduces the extent to which populations may differ; restricted gene flow allows enhanced differences. Populations of some geographically widespread species may differ so much that they may be unable to interbreed.
- If barriers to breeding accompany differences in visible traits, the species may be divided into **subspecies**.
- Continuous geographic variation is usually described in terms of **clines** (character gradients across a map).
- Geographic variation is the first step in species formation.

Key 63 Evolution of species

OVERVIEW *Species are evolutionary units within which gene flow occurs. Natural populations belong to the same species only if they can interbreed and leave fertile offspring. Different species are reproductively isolated from each other. Most new species originate geographically.*

Importance of species: Species characteristics are passed from parents to offspring. Species were long considered real and important because these characteristics were unchanging (unevolving).

Morphological (typological) species definition: Each species is defined by characteristics considered "essential" or "typical."

Biological species definition: Species are groups of interbreeding populations that are reproductively isolated from other species.

Reproductive isolating mechanisms: These can act prior to mating:
- **Ecological isolation:** Potential mates do not meet because they live in different habitats or breed at different times or seasons.
- **Behavioral isolation:** Mating calls or mating rituals differ.
- **Mechanical isolation:** "Lock-and-key" mismatch of genitalia.
or after mating:
- **Gametic mortality:** Gametes die in female reproductive tract.
- **Zygotic mortality:** Fertilized eggs fail to divide properly.
- **Embryonic or larval mortality:** They die prematurely.
- **Hybrid inviability:** Hybrids never reach reproductive age.
- **Hybrid sterility:** Hybrids cannot reproduce, as in mules.
- **F_2 breakdown:** Offspring of hybrids are inviable.

Geographic speciation: Most speciation occurs geographically:
- A species develops geographic variation over its range.
- Geographic barriers prevent contact between populations.
- Reproductive isolating mechanisms may now evolve.
- Geographic isolation ends, with two possible outcomes:
    1. No reproductive isolation: still a single species.
    2. Reproductive isolation is effective: two species now exist; selection will enhance differences between them.

Other models of speciation (sometimes controversial):
- By doubling of chromosome number in plants.
- "Semigeographic" or "parapatric": adjacent ranges may touch.
- "Sympatric" and "stasipatric": no geographic separation.

Key 64 Macroevolution

OVERVIEW *Evolution above the species level (**macro-evolution**) consists of two distinct processes: changes within a lineage (anagenesis), and the branching of lineages (cladogenesis). Evolutionary trends are adaptive and are usually opportunistic, following no plan or goal but rather taking the path of least resistance. Cladogenesis fills the biosphere with an ever-increasing number of species, arranged into classifications which reflect descent. Species that fail to adapt to changing conditions become extinct.*

Lineage: An ancestor-to-descendent sequence of species.

Trend: Continued morphological change within a lineage.

Parallelism: Independent occurrence of the same or similar trends in different lineages.

Convergence: Similar adaptations in unrelated lineages.

Cladogenesis: The branching of lineages by speciation.

Anagenesis: Evolution within a lineage, between branching points.

Evidence for the adaptiveness of trends:
- Trends often persist for a long time.
- Parallel trends often occur independently.
- Evolutionary rates vary: trends speed up or slow down; they may even stop altogether or reverse direction.
- Trends in different characters do not always go together but occur independently (**mosaic evolution**) and at different rates and times. For this reason, transitional species (like *Archaeopteryx*) are a mosaic of primitive and advanced features mixed together.

Opportunism: Evolution follows no plan or goal but instead takes the path of least resistance:
- Cladogenesis fills the biosphere with more and more species.
- Diversity among the descendents of a single species (**adaptive radiation**) often results.
- Functional problems are often solved differently in different lineages (**multiple solutions**, such as diversity among eyes).
- The same trend often occurs repeatedly (**iterative evolution**).
- Convergence (and its imperfections) shows that similar adaptive opportunities may arise independently more than once.

- Organs that change function usually serve both old and new functions simultaneously during the transition.

Rates and modes of evolution: Evolutionary rates may measure either anagenesis, cladogenesis, or both; rates calculated in different ways are usually not comparable.
- Since Darwin, most evolutionists have viewed evolution as a continuous, gradual process.
- Many scientists now view evolution as a series of steady equilibria punctuated by infrequent episodes of very rapid change (the **punctuated equilibrium** theory).

Results of evolution: The results of macroevolution can be seen in the diversity among species that is reflected in their anatomical structure and in our classifications. These results include:
- **Adaptations:** Features which help organisms cope with and exploit their environments.
- **Analogy:** Similarities among species resulting from adaptation to similar functional requirements.
- **Homology:** Deep-seated resemblance reflecting common ancestry.
- Evolution results in classifications that contain groups within groups (Key 66); these groups have always been considered "natural," even by pre-evolutionary taxonomists. Whenever new technology allows new types of variation to be studied, most variation follows the patterns of groups that were recognized beforehand on the basis of other criteria.

Extinction: Species that cannot adapt to change become extinct.
- Extinction may occur either early or late in the history of a group.
- Extinction may occur at times of either low or high diversity.
- Some paleontologists believe that rates of speciation and of extinction are usually equal.

Key 65 Fossils and paleontology

OVERVIEW *The geological time scale divides the last 600 million years into 12 periods. Fossils differ in the degree to which smaller structural details are preserved and in the degree of chemical alteration of the original material. The fossil record allows us to test various evolutionary theories against the actual long-range history of life on Earth.*

Geologic time scale: The Earth is about 5.5 billion years old, but only the last 600 million years is well documented by fossils.

Precambrian Era (up to about 600 million years ago): Includes the earliest fossils, about 5.2 billion years old. Precambrian fossils are very rare; most are microscopic fossils of procaryotic organisms.

Paleozoic Era (up to about 200 million years ago): The time when invertebrates dominated the oceans and when fishes, insects, land plants, and amphibians first flourished. Divided into seven periods:
- **Cambrian** (oldest)
- **Ordovician**
- **Silurian**
- **Devonian**
- **Mississippian (= lower Carboniferous)**
- **Pennsylvanian (= upper Carboniferous)**
- **Permian** (most recent).

Mesozoic Era (up to about 65 million years ago): Sometimes called the **"Age of Reptiles"** because dinosaurs and other large reptiles dominated the land while marine reptiles (and ammonoid molluscs) flourished in the seas. Mesozoic time is divided into three periods:
- **Triassic** (oldest)
- **Jurassic**
- **Cretaceous** (most recent)

Cenozoic Era ("Age of Mammals"): The last 65 million years. It is divided into two periods:
- **Tertiary** (from 65 to 2 million years ago)
- **Quaternary** (the last 2 million years), including Pleistocene and Recent epochs.

Paleontology: The study of fossils.

Fossils: Remains or other evidence of the life of past geologic ages.

- **Fossils containing original material:**
    1. **Unaltered remains:** *Example:* frozen mammoths.
    2. **Compressions:** Flattened and dehydrated, but unaltered otherwise, with cellular details often preserved.
- **Replacement fossils** (with original material largely replaced):
    1. **Permineralization, impregnation, and embedding:** Gradual *addition* of minerals by ground water, preserving many internal details.
    2. **Carbonization:** Volatile components lost, leaving carbonized skeleton only.
    3. **Mineralization:** Complete replacement of original material by minerals.
- **Casts and molds:** Impressions in fine-grained sediments, preserving only surface shapes. **Casts** are solid objects; **molds** are hollow.
- **Trace fossils:** Tracks, trails, footprints, burrows, and other traces of activity. *Examples:* **amber** (fossil resin), **coprolites** (fossil dung).

Lessons learned from studying the fossil record: The fossil record can be used to test various theories against the actual record of life on Earth. No proposed theory or evolutionary mechanism is acceptable if it conflicts with this historical record.

- The fossil record shows a process of branching and diversification, not just a linear sequence. "Evolution is a bush, not a ladder."
- **Cope's rule:** Size increases frequently and decreases much less often.
- **Williston's rule:** Repeated parts (like multiple legs or segments) often become less numerous and more different from one another.
- **Dollo's law:** Like other historical processes, evolution never repeats itself exactly. Because of probability considerations, small, simple changes may reverse, but larger and more complex changes never do. Evolution is thus constrained (limited) by its own history.

Theme 10 ORGANIC DIVERSITY

*B*iological diversity is expressed by arranging organisms into kingdoms, phyla, classes, orders, families, genera, and species. Evolutionary relationships responsible for these arrangements are often depicted in family trees.

We now believe that life originated on Earth in a reducing atmosphere containing no free oxygen. The first organisms had simple cells with no nuclei and got their energy from energy-rich compounds in their environment. Photosynthesis evolved later. Organisms with true nuclei (eucaryotes) came much later and probably originated by symbiosis. An important milestone in plant evolution was the enclosing of the egg cell in nonreproductive tissue, as in mosses. Another important milestone was the evolution of vascular tissue that could support a larger plant and circulate fluids within it. Important early milestones in animal evolution include the origin of tissue layers in the Cnidaria, the origin in flatworms of a mesoderm layer and of bilateral symmetry, and the independent origin of body cavities in several groups. Animals with stiff notochords (chordates) include the fishes, amphibians, reptiles, birds, and mammals.

Key 66 Diversity and its classification

OVERVIEW *Organic diversity is described by grouping species into genera, genera into families, families into orders, orders into classes, classes into phyla, and phyla into kingdoms. These groups reflect evolutionary history and common ancestry as much as possible.*

Binomial nomenclature: Each species has a two-word name. The first word (capitalized) is the name of the genus; the second word (lower case) is the name of the species. *Example: Homo sapiens.*

The Linnaean system: Uses binomial nomenclature throughout. Species are grouped into genera and genera into higher groups. Any one of these groups, at any level, is called a **taxon** (plural, taxa). The complete Linnaean hierarchy (ranking) of groups is as follows:
Kingdom (the most inclusive group)
  Phylum (plural, phyla; sometimes called a "division" in plants)
   Class
    Order
     Family
      Genus (plural, genera)
       Species (same spelling in singular and plural).
Extra ranks are added to this hierarchy as needed, such as subphylum (just below phylum) or superfamily (just above family). *Example:* Humans belong to kingdom Animalia, phylum Chordata, class Mammalia, order Primates, family Hominidae, genus *Homo*, species *Homo sapiens.*

Evolutionary classification: Biological classification is believed to reflect the results of a branching evolutionary process. Insofar as possible, our classifications should be genealogical. Each taxon should ideally represent one branch of the evolutionary tree, with the smaller included taxa representing its sub-branches.

Five kingdoms: Most biologists now arrange organisms into five kingdoms:
- **Monera:** Procaryotic organisms (bacteria and blue-green bacteria).
- **Protista:** Simple eucaryotic organisms, generally one-celled.
- **Fungi (Mycota):** Fungi, with cell walls but no plastids.
- **Plantae:** Plants, containing plastids and chlorophyll.
- **Animalia:** Multicellular animals, developing from blastulas.
  Some biologists recognize Archaebacteria as a sixth kingdom.

Key 67   Family trees and taxonomic theory

OVERVIEW   *The aim of phylogenetics is to reconstruct family trees and to base classifications upon them. Rival theories of classification include phenetics (basing classifications on resemblance only) or cladistics (basing classifications on branching sequences only). Evolutionary classifications are based on all available evidence.*

Phylogeny: A family tree of species.

Phylogenetics: The study of family trees.
- Phylogenetic methods use both the fossil record and resemblances among living organisms as evidence to reconstruct phylogenies. Organisms sharing many similarities are considered to be descendents of a common ancestor that also shared these similarities. When conflicting evidence arises from different characters, further study is undertaken to see if all similarities are as real as they look, or if some similarities could have evolved by convergence.
- An important task in phylogenetics is distinguishing homology, which reflects common ancestry, from analogy or convergence (Key 64).
- The aim of classification based on phylogenetics is to group together those species that derive their similarities from a common ancestor. This means that, insofar as possible, each taxon should be made **monophyletic** by including the common ancestor within the taxon.

Taxonomy: Is the theory behind the making of classifications. All theories of taxonomy, except phenetics, are based on phylogenetics.

Phenetic taxonomy: Bases classifications entirely upon degrees of resemblance, ignoring all attempts to reconstruct phylogenies. Phenetic taxonomy is now generally in disfavor because it does not distinguish convergence from other sources of resemblance.

Cladistic taxonomy: Bases classifications on the geometry of branching alone, ignoring such matters as the length of twigs, the diversity of branchings, or the degree of change along lineages.

Eclectic (evolutionary) taxonomy: Bases classifications on all the available evidence, including both sequences of branching and degrees of resemblance among descendents.

Key 68 The origin of life

OVERVIEW *Under present conditions, Louis Pasteur demonstrated that life can come only from preexisting life. Modern ideas on the origin of life follow Oparin's suggestion that life originated in a reducing atmosphere consisting of hydrogen, methane, ammonia, and water vapor. Miller showed that amino acids could arise spontaneously in such an atmosphere. Simple proteins probably arose from such amino acids. Self-perpetuating systems were selected and replicated while other systems unraveled. The origins of DNA replication and modern-style protein synthesis are currently the subject of several competing theories.*

Spontaneous generation: Theory prevailing before Pasteur, that life could easily and spontaneously arise from nonlife.

Francesco Redi (1600s): Disproved the spontaneous generation of flies; showed that the larvae came from tiny eggs, not from rotting meat.

Invention of the microscope (around 1700): Led to the discovery of bacteria. Early experiments, flawed by poor sterilization, seemed to show that bacteria could arise from nonliving matter.

Louis Pasteur (1860s): Perfected sterilization techniques and reenacted all earlier experiments. He proved that properly sterilized broth would remain sterile if bacteria were excluded, but that ordinary air contained bacteria that could contaminate the broth unless precautions were taken. This led to the **theory of biogenesis** — life can originate only from preexisting life.

Alexander Oparin (1930s): Proposed that the origin of life was impossible under present conditions, but that life originated under very different conditions on the primitive Earth **(primary abiogenesis)**. He postulated that life could originate only in a hydrogen-rich **reducing atmosphere**, and he proposed that Earth's original atmosphere contained **hydrogen (H_2), methane (CH_4), water vapor (H_2O), and ammonia (NH_3)**. J.B.S. Haldane proposed a similar theory independently, but most scientists ignored these ideas until the 1950s.

S.L. Miller (1950s): Tested Oparin's ideas by combining H_2, CH_4, H_2O, and NH_3 in a sterile apparatus into which he could introduce a spark to simulate lightning. After circulating this mixture for several days, he analyzed the products and found many amino acids and a few oligopeptides.

Chemical evolution and the origin of life: Current ideas about the origin of life are based on the Oparin-Haldane theory of **chemical evolution**, in which life arose gradually in a reducing atmosphere.

- The solar system probably formed from a swirling nebula with the sun in the center and planets on the periphery.
- Amino acids probably originated in a manner similar to the reactions of Miller's experiment. The compounds dissolved in the primitive ponds and oceans, forming a "hot, dilute soup."
- Proteins and DNA can form as **polymers** by linking smaller units together, but not until the smaller units are concentrated. Several concentration mechanisms (tidal pools, crystal surfaces, bubble-like droplets, etc.) have been suggested.
- Molecules made without life are usually symmetrical or have equal proportions of right-handed and left-handed forms, but biological systems contain mostly asymmetrical molecules. Amino acids made by organisms are mostly of the L- (left-handed) form, but experiments like Miller's give right and left-handed amino acids in equal proportions. Molecular asymmetry is an important property of life, but we don't know exactly when or how it arose.
- At some point, early biological systems probably formed tiny droplets with lipid or protein membrane-like surfaces. Different authorities have imagined different kinds of droplets, calling them "coacervates," "microspheres," "protobionts," etc. Once these droplets formed, their contents could reach concentrations very different from those prevailing outside or from one another (they had individuality). Some were surely more stable than others, and were favored by **"protoselection,"** especially if they could increase in size and fragment into smaller droplets, a primitive form of reproduction.
- Protein synthesis was surely much simpler originally than it is now and was probably less reliable in perpetuating sameness. Enzyme activity may have originated by chance. The origins of DNA replication are obscure. A few biochemists believe that DNA replication came before protein synthesis, but most favor the "protein first" viewpoint, in which RNA and DNA were initially selected for their role in making protein synthesis more reliable.

Exobiology: The search for life elsewhere, outside planet Earth. To date, much evidence exists for Miller-style synthesis of amino acids, nitrogen bases, and other compounds elsewhere in our solar system. No firm evidence has yet been found that life formed anywhere except on Earth, but many scientists think such origins very probable.

Key 69 Procaryotes

OVERVIEW *Procaryotes are one-celled organisms whose cells lack a true (membrane-bounded) nucleus and other eucaryotic organelles. Procaryotes include the Archaebacteria, true bacteria, and Cyanobacteria.*

Procaryotic cells: Cells without true nuclei, lacking many other structures found in eucaryotic cells (Key 11).

Procaryotic cell walls: Contain substances like muramic acid, absent in eucaryotes.

Chemical diversity: Procaryotes have greater chemical diversity than eucaryotes. They can subsist on a greater variety of foodstuffs, have a greater range of chemical substances that can be tolerated, and can subsist in a variety of atmospheres, both with and without oxygen.

Procaryotic chromosomes: Generally arranged in a single circular loop containing DNA but no protein. Partial recombination may occur during conjugation. Most procaryotes also have small chromosome fragments that can detach from the main chromosome and exist separately for long periods as **plasmids**, small circular samples of DNA similar to certain viruses.

Archaebacteria: A small group of strict anaerobes (killed by oxygen) that include the methane-producers (methanogens), the extreme halophiles, and the extreme thermophiles. Their RNA sequences have only minimal homology to the RNA of the other procaryotic or eucaryotic organisms, and the cell walls are also unique.

True bacteria: The majority of procaryotes, with RNA sequences homologous to those of eucaryotes and Cyanobacteria (but not Archaebacteria). Most are heterotrophs (Key 57). The few autotrophs use a variety of energy sources, but none contains chlorophyll *a* and none can split water in the Hill reaction (Key 17).

Cyanobacteria (=Cyanophyta, blue-green bacteria, or blue-green algae): Are similar to bacteria in structure and their RNA sequences are homologous. All are oxygen-tolerant autotrophs that can use sunlight for energy and CO_2 as a carbon source. They contain chlorophyll *a* and can split water in the Hill reaction (Key 17).

Key 70 Origin of eucaryotes

OVERVIEW *Eucaryotic cells have nuclei surrounded by nuclear membranes. The several chromosomes each contain protein as well as DNA. The cytoplasm of eucaryotic cells contains many organelles not found in procaryotic cells. Much evidence indicates that eucaryotic cells originated by symbiosis and particularly that mitochondria and plastids were once separate organisms with their own DNA.*

Eucaryotic cells: Cells with true nuclei, each containing a nucleolus and surrounded by a nuclear envelope.

- Chromosomes are usually separate, multiple, and linear; each contains protein as well as DNA.
- Cytoplasm contains contractile proteins (actin, myosin), making possible cytoplasmic streaming (cyclosis), amoeboid locomotion using pseudopods, and ingestion of food by phagocytosis.
- Many types of cytoplasmic organelles are present (Key 11), including membrane organelles (like the endoplasmic reticulum) and "9 + 2" groupings of microtubules. Mitochondria (and plastids in plant cells) contain their own DNA.

Theory of endosymbiosis: Most biologists believe that eucaryotic cells originated when small, energy-producing procaryotic cells lived inside larger cells and became mitochondria by intracellular symbiosis (**endosymbiosis**). Plastids may have arisen the same way. Similar origins for other cell structures have been proposed but are less widely accepted. Perhaps host cells originally phagocytized the energy-producing mitochondria; then natural selection favored those host cells that maintained the mitochondria as an energy source instead of digesting them.

- Evidence for endosymbiosis comes from the fact that mitochondria and plastids have two membranes: the outer one resembles the membranes of the eucaryotic host cell, but the inner membrane resembles procaryotic cell membranes instead.
- Mitochondria and plastids possess their own DNA and are self-replicating.
- Bacteria are known whose metabolic abilities are similar to those of the postulated host cell, while other bacteria have all the enzymes of the Krebs cycle and are thus comparable in ability to mitochondria.

Key 71 Protozoa

OVERVIEW *The one-celled non-photosynthetic eucaryotes (Protozoa) are probably the group from which all other eucaryotes are derived. They possess a diversity in locomotor adaptations: the Sarcodina move mostly by amoeboid locomotion using pseudopods; the Mastigophora move by flagellae; and the Ciliata move by cilia. The non-motile Sporozoa reproduce by spores and may be close to the ancestry of fungi.*

Locomotor adaptations: Distinguish the various types of Protozoa:
- **Sarcodina:** Move by **amoeboid locomotion** using protoplasmic extensions called **pseudopods**.
- **Mastigophora:** Move by means of a whip-like **flagella**.
- **Ciliata:** Move by means of cilia which cover the body surface.
- **Sporozoa:** Are non-motile and reproduce by spores.

Phylum Sarcodina: The largest and most diverse group of Protozoa, including the amoebas. Locomotion uses protoplasmic extensions called **pseudopods**; the body changes shape continually as it moves. Most Sarcodina are predators, engulfing their prey by phagocytosis.

Phylum Mastigophora (Flagellata): Flagellated unicells that move by beating a long, whip-like **flagella** with a "9 + 2" grouping of microtubules. Some species show both flagellar and amoeboid locomotion, giving evidence that the Mastigophora and Sarcodina are related.

Phylum Sporozoa: A group of non-motile unicells, including the malarial parasite, *Plasmodium*. Because the Sporozoa reproduce by spores, they may be close to the ancestry of the slime molds and fungi.

Phylum Ciliata (Ciliophora): Cells covered with **cilia** organized into a continuous layer or **pellicle**. Beating of the cilia control locomotion and also feeding. The plasma membrane forms a food vacuole by budding inward, a process similar to phagocytosis. The Ciliata are unusual in having two nuclei in each cell, which makes for a very complex process of sexual conjugation, so different from other Protozoa that many biologists believe the Ciliata to have evolved independently. *Paramecium* is a commonly studied ciliate.

Plant-like Protista: Pyrrophyta, Chrysophyta, and certain other photosynthesizers are sometimes called alga-like Protista, but they all possess plastids and plant-like pigments, and so are often treated as plants (Key 73).

Key 72 Life cycles

OVERVIEW *Many organisms have life cycles in which diploid stages (sporophytes) alternate with haploid stages (gametophytes)—an **alternation of generations** in which meiosis marks the beginning of the gametophyte stage and fertilization marks the beginning of the sporophyte stage. The sporophyte generation has assumed greater and greater importance throughout much of plant evolution.*

Gametophyte: Any multicellular haploid body.

Gametophyte phase or generation: Haploid portion of life cycle, from meiosis to fertilization.

Sporophyte: Any multicellular diploid body.

Sporophyte phase or generation: Diploid portion of life cycle, from fertilization until the next meiosis.
  * Either sporophyte or gametophyte phase (or both) can be conspicuous. Most microorganisms have dominant haploid stages, while higher plants and most animals have dominant diploid stages.

Evolution of life cycles:
  * The unicellular green alga *Chlamydomonas* as a dominant haploid phase. The diploid zygote exists only briefly before it undergoes meiosis and releases new haploids. Gametes of opposite (+ and −) types look the same, a condition called **isogamy**.
  * Life cycles vary greatly among other algae. Gametes are often dissimilar **(anisogamy)**. Often, the male gametes are small and motile, while eggs are much larger and nonmotile **(oögamy)**.
  * Moss plants are gametophytes, which contain sex organs and produce gametes. Sporophytes grow as virtual parasites on the gametophytes. Meiosis produces haploid spores, which grow into new gametophytes.
  * The small fern gametophyte resembles a heart-shaped leaf; sex organs are produced on its surface. Fertilization produces a sporophyte, which is the conspicuous fern. Meiosis produces haploid spores, which grow into new gametophytes.
  * In seed plants, the conspicuous plant is always a sporophyte. Angiosperms produce microscopic female gametophytes of only a few cells within the ovaries of the flower. Male gametophytes are even smaller and are contained in the pollen grains. The seed that results after fertilization contains a young sporophyte.

Key 73 Algae

OVERVIEW *Eucaryotic algae include the green algae (Chlorophyta), brown algae (Phaeophyta), red algae (Rhodophyta), and several groups of microscopic algae. Some biologists treat the algae as plants; others treat them as Protista.*

Green algae (Chlorophyta): Characteristic pigments, similar to those of higher plants, include chlorophylls *a* and *b*, xanthophylls, and α- and β-carotenes (or γ- and β-carotenes in one group). Starch is the main storage product, as in higher plants. Main cell wall constituents are cellulose and pectin. Chlorophyta are ecologically dominant in freshwater ecosystems, but many are also marine. Of all groups of algae, the Chlorophyta are most similar to the higher plants.

Euglenoids (Euglenophyta): One-celled, many pigments similar to green algae. No cell walls, except for some strips of protein. Can use photosynthesis or animal-like ingestion, depending on light levels.

Brown algae (Phaeophyta): Largest of all algae, ecologically dominant in temperate and colder marine waters. Structurally most complex algae, with frequent division into a holdfast, a stalk or stipe, and a leaflike blade. Pigments include chlorophylls *a* and *c*, plus several unique xanthins. Many unique storage products and cell wall components.

Dinoflagellates (Pyrrophyta): Small algae, mostly marine and planktonic. Pigments include chlorophylls *a* and *c*, plus some unique pigments. Two flagellae, arranged at right angles to one another. Mitosis unusual: no histones, no centrioles, no spindle fibers; nucleolus and nuclear membrane remain visible throughout mitosis.

Golden-brown algae and diatoms (Chrysophyta): Diatoms are especially abundant in plankton and are the major primary producers in most marine ecosystems. Some pigments and storage products are unique; others show similarities to brown algae.

Yellow-green algae (Xanthophyta): A group related to the Chrysophyta, but distinguished by pigments that include chlorophylls *a* and *e*.

Red algae (Rhodophyta): Algae with chlorophylls *a* and *d*, plus other pigments resembling those of Cyanobacteria. No "9 + 2" organelles (centrioles or flagellae); no motile cells, not even gametes. Unique storage products include "floridean starch" in cell walls. Ecologically dominant in tropical marine waters.

Key 74 Fungi

OVERVIEW *The **Mycota** or **Fungi** are non-photosynthetic organisms adapted to absorptive nutrition. Slime molds have motile, unicellular vegetative stages, while true fungi usually form branched filaments that invade dead or decaying material. All fungi form spores; the various groups are distinguished by the methods of spore formation.*

General characteristics of fungi: Plastids and chlorophyll are absent. Cell walls are not made of cellulose. Cell membranes occasionally break down to form binucleated cells or multinucleated aggregates. Reproductive structures vary, but spores are always produced. Nutrition is usually absorptive; many fungi live on dead or decaying matter **(saprophytic)**, but some are parasitic instead. Fungi are important as decay organisms in freshwater and terrestrial ecosystems. Most prefer moist conditions for optimal growth.

Slime molds: Fungi whose unicellular vegetative stages are either amoeboid or flagellated and resemble Protozoa. All types have a multinucleated or multicellular creeping stage that forms spore-producing bodies. Each spore develops into a new vegetative cell.

True fungi (Eumycota): Fungi whose vegetative structure typically consists of a series of branching filaments **(hyphae)** forming a tuft **(mycelium)**.

Fungi with aquatic stages: Four primitive classes of fungi still reproduce by flagellated cells (zoospores).

Zygomycetes (black bread molds, etc.): Reproduce by conjugation of hyphae that come together and form nonmotile spores.

Ascomycetes (yeasts, common bread molds, cup fungi, truffles, etc.): Spores are produced, 4 or 8 at a time, in sacs (asci).

Basidiomycetes (mushrooms, puffballs, rusts, smuts, etc.): Spores are produced, usually 4 at a time, at the tips of club-like organs (basidia).

Deuteromycetes ("fungi imperfecti"): Fungi with no known sexual stages.

Lichens: Very intimate symbiotic associations of fungi with either algae or cyanobacteria. The fungus absorbs and retains sufficient moisture for both partners; the green partner photosynthesizes and provides food. Lichens are often the first colonizers of bare rock surfaces.

Key 75 Bryophytes

OVERVIEW *Bryophytes develop from multicellular embryos containing sterile, nonreproductive cells that surround and protect the zygote. However, they lack the vascular tissues which allow higher plants to grow very tall. All bryophytes have alternation of generations with the gametophyte stage dominant. The two major types of bryophytes are the liverworts (with the related hornworts) and the mosses.*

Embryophyta: Bryophytes and all higher plants are sometimes called **Embryophyta** because they develop from **embryos** in which the zygote is surrounded by a protective layer of sterile, nonreproductive cells. Multicellular sex organs are present.

Bryophytes: Embryophyta without vascular tissues (xylem, phloem). Because they lack vascular tissues or true roots, bryophytes cannot be anchored strongly in the soil or grow very tall. Neither water nor nutrients can be transported from one plant part to another, except by diffusion. This restricts bryophytes to small sizes and to moist habitats; it also means that all parts of the plant must carry out their own photosynthesis. Bryophytes probably evolved from green algae. All bryophytes have a well-marked alternation of generations (Key 72).

Liverworts and hornworts: Bryophytes whose gametophytes are mostly flat-lying plants with distinct upper and lower surfaces. Single-celled absorptive **rhizoids** grow from the lower surface. Sporophytes vary, but are generally simple.

Mosses: Bryophytes whose gametophytes usually have an erect, stem-like portion surrounded with leafy extensions arranged in a circular pattern. Absorptive rhizoids are often multicellular, with cross-walls. Sporophytes are typically more complex than in liverworts, with a spore-bearing capsule supported by a stalk.

Key 76 Vascular plants without seeds

OVERVIEW *Vascular plants are called **tracheophytes**. Vascular tissue provides support that holds plants erect and allows them to grow much taller; it also allows plants to transport materials from one part to another.*

Vascular tissues: Plant tissues that conduct fluids through cells with stiff cell walls. The stiff cell walls (wood) allow plants to grow taller and erect. By conducting fluids, vascular tissues allow above-ground plant parts to receive water and nutrients absorbed by the roots, and they allow the parts that do not photosynthesize to receive sugars and other products from the green, photosynthesizing parts.

Early vascular plants: The earliest vascular plants were Silurian plants like *Rhynia* and *Asteroxylon,* living in moist, swampy places. **Dichotomous** (two-fold) **branching** characterized all plant parts. Much of each plant grew out horizontally, but some parts turned upward and grew erect. No true leaves were present; stems were green and photosynthetic.

Psilophyta: Includes only a few living genera *(Psilotum, Tmesipteris),* with spore-forming structures **(sporangia)** terminal in position. Stems are green and photosynthetic; no true leaves or roots are present. Stomates (Key 38) are distributed over the outside surface of the stems.

Lepidophyta (lycopods): Club mosses and their relatives. True roots are present and are dichotomously branched. Leaves (mostly small) are **microphylls**—each has a single vein in the middle, and the vascular bundle is not interrupted where this vein arises. Some leaves bear reproductive sporangia in the angle of attachment (axillary position). Modern lycopods are small plants, but some Carboniferous lycopods grew to tree-like heights.

Arthrophyta or Sphenopsida: Horsetails, with spore-forming parts grouped in a cone-like structure at the top; spores are hidden beneath scale-like **sporophylls**. Small leaves are arranged in tiers or whorls; true roots are present. *Equisetum* is the only living genus.

Pterophyta: Ferns and fern-like plants, with true roots. Leaves are **megaphylls**—each has many branching veins, and the vascular bundle is interrupted by a **leaf gap** where the main vein arises. Leaves carry sporangia on their lower surfaces or their margins. Life cycle (Key 72) with dominant sporophyte.

Key 77   Vascular plants:
Gymnosperms

OVERVIEW *Vascular plants with naked seeds are placed in five or more phyla (divisions): seed ferns (Pteridospermophyta), cycads (Cycadophyta), ginkgos (Ginkgophyta), conifers (Coniferophyta), and Gnetophyta.*

Gymnosperms: Vascular plants with naked seeds. They all possess megaphylls (leaves with branched veins whose bases interrupt the vascular bundle to form a leaf gap). They share with angiosperms the presence of a **seed**, an easily dispersed structure developed from the zygote and enclosing the embryonic sporophyte. They differ from angiosperms in that the seed is not enclosed within an ovule, but lies naked on the surface of a reduced, scale-like leaf.

Seed ferns (Pteridospermophyta): Extinct plants with large, fern-like leaves, but reproducing by seeds; occasionally growing to the height of small trees. Devonian to Jurassic in age; dominant during the Carboniferous. Believed close to the ancestry of other seed plants.

Cycads (Cycadophyta): Short, thick-stemmed plants with a crown of large fern-like or palm-like leaves. Seeds borne together in a structure resembling a large pine cone. Flourished during the Mesozoic era; only a few tropical and subtropical genera persist today.

Ginkgos (Ginkgophyta): A mostly Mesozoic group with one living species (*Ginkgo biloba,* an ornamental tree with fan-shaped leaves).

Conifers (Coniferophyta): The most familiar and economically important gymnosperms, including pines, spruces, firs, etc. Leaves are typically scale-like or needle-like, with reduced surface area. Seeds borne in cone-like aggregates.

Gnetophyta: This group contains only three living genera *(Gnetum, Ephedra, Welwitschia),* which differ greatly. All share a partially enclosed type of seed that approaches the angiosperm condition but was probably derived independently.

Key 78 Vascular plants: Angiosperms

OVERVIEW *Angiosperms (flowering plants) have seeds enclosed in an* **ovary**. *Flowers include ovaries and surrounding structures. Ripened ovaries make up a* **fruit**. *Roots, stems, and leaves reach maximum complexity and diversity within the angiosperms. Angiosperms are divided into "dicots" (Dicotyledonae) and "monocots" (Monocotyledonae).*

Angiosperms: Flowering plants, in which seeds are enclosed in a protective **ovary**. Many flowers are fertilized by insects. Angiosperms probably evolved in response to selection by insects.

Flower: An ovary, together with surrounding reproductive structures:
- **Petals** and **sepals:** Surround and protect the other flower parts, especially in undeveloped flowers. In many species, these parts may have colors and/or odors that attract insects or other species that pollinate the plant or disperse its seeds.
- **Male flower parts:** Include **stamens**, each made of a **filament** and a pollen-producing **anther**. **Pollen grains** contain male gametophytes.
- **Female flower parts:** Include **stigma** (whose sticky surface catches pollen), **style** (a stalk-like part supporting the stigma), and **ovary**, enclosing one or more **ovules** within modified leaves called **carpels**. Each ovule develops into an 8-celled gametophyte containing 1 egg, 2 polar nuclei, and 5 other cells.
- **"Complete" flowers:** Have all the parts listed above, but many species have separate male flowers and female flowers.

Fertilization: First, a long, thin **pollen tube** grows out from the pollen grain down the style. Two nuclei (**tube nucleus** and **generative nucleus**) migrate down the pollen tube as it grows. The generative nucleus then divides into two **sperm cells**: one fertilizes the egg (**syngamy**, or **true fertilization**); the other fuses with the 2 polar nuclei to form a **triploid (3N) endosperm**, containing stored food that supplies the embryo later. This so-called **double fertilization** is characteristic of all angiosperms. After fertilization, the new embryo and its endosperm and protective coverings make up a **seed**.

Fruits: One or more ripened ovaries together constitute a **fruit**, enclosing one to many seeds. Many fruits are eaten (and the seeds dispersed) by animals; other seeds are dispersed by wind, etc.

Roots: Absorptive parts, often underground. They contain, from the outside inward:
- **Epidermis:** From which absorptive **root hairs** develop.
- **Cortex:** Often a thick layer.
- **Endodermis:** Containing a waterproof **Casparian strip**.
- **Vascular bundles:** With xylem, phloem, and cambium surrounded by a **pericycle**.

Stems: Supportive structures which contain, from the outside inward:
- **Epidermis**
- **Cortex**, or **bark**
- **Vascular bundles**, each containing phloem and xylem (Key 36) along with some persistently embryonic tissue called **cambium**.
- **Pith** (not always present) in the innermost part of the stem.

Leaves: The major photosynthetic organs. Leaf structure is described in Key 36.

Class Dicotyledonae ("dicots"): Angiosperms with a seed containing two "seed-leaves" **(cotyledons)** in which food is stored (the two halves of a dried peanut are a familiar example).
- Flower parts are usually in multiples of four or five.
- Veins in leaves and petals branch to form net-like patterns.
- Vascular bundles usually arranged in a circular ring.
- Includes the majority of angiosperms: buttercups (considered primitive), roses, beans (and other legumes), daisies, oaks, maples, apples, peaches, oranges, melons, and many others.

Class Monocotyledonae ("monocots"): Angiosperms with a seed containing only one "seed-leaf" (cotyledon) in which food is stored (such as a kernel of corn, which cannot be divided into halves).
- Flower parts are usually in multiples of three or six.
- Leaves (and petals) have parallel veins that branch only occasionally.
- Vascular bundles usually scattered throughout cross-section of stem.
- Includes the more advanced angiosperms: orchids, lilies, palms, and grasses (including wheat, corn, and other cereal grains.)

Key 79 Sponges and cnidaria

OVERVIEW *Sponges (phylum Porifera) are unsymmetrical or radially symmetrical animals with separate cell types but no distinct tissues. Their bodies have many pores and sharp protective spicules. Coelenterates (phylum Cnidaria) are radially symmetrical animals with two tissue layers: an outside layer (ectoderm) and an inside layer (endoderm) surrounding an all-purpose gastrovascular cavity.*

Phylum Porifera (sponges): Aquatic animals, with radial symmetry or irregular shapes. Water passes through many incurrent **pores**, which often lead to a central cavity, then exits, often by an excurrent opening or **osculum**. No distinct tissues, but many cell types:

- **Epidermal cells (pinacocytes):** Outside lining.
- **Porocytes:** Barrel-shaped pore cells.
- **Choanocytes:** Flagellated **collar-cells**; keep water flowing.
- **Amoebocytes:** Several kinds of amoeboid cells; some secrete spicules.
- **Spicules:** Sharp needles or more complex shapes embedded within sponges, functional in support and as a defense against predators. May be composed of silica, calcite, or horny protein; differences in spicule shape and composition are used in classification.

Phylum Cnidaria (Coelenterates): Aquatic animals with two body layers (outer **ectoderm** and inner **endoderm**) separated by a jelly-like **mesoglea** and an all-purpose **gastrovascular cavity** with a single opening **(mouth)**. **Tentacles** surround mouth and have stinging cells **(cnidocytes)** containing stingers **(nematocysts)**. Two major body forms: **polyp** (with mouth directed upward, mesoglea thin; animal often attached) and **medusa** (free-swimming "jellyfish" with thick mesoglea and mouth directed downward).

- Class **Hydrozoa**: Life cycle includes both asexual polyps and sexually reproducing medusae (usually small). Solitary or colonial; some colonies have many types of individuals interconnected.
- Class **Scyphozoa**: Solitary "jellyfish," with dominant medusa stage.
- Class **Anthozoa**: The largest class, including sea anemones and corals. Polyp stage dominant; no medusa. Mouth extends inward to form a tubular **pharynx**. Solitary or colonial.

Phylum Ctenophora ("comb jellies"): A small group of marine animals with **biradial** symmetry (like a two-armed pinwheel). Two large tentacles and 8 comb-like rows of cilia.

Key 80   Flatworms and related
wormlike animals

OVERVIEW   *Flatworms and nearly all other animals from here on are bilaterally symmetrical (right and left halves are mirror images). The front end of such animals usually forms a distinct* **head**. *However, flatworms still have a single all-purpose cavity with only one opening.*

Bilateral symmetry: Right and left halves are mirror images of each other.

Phylum Platyhelminthes: Flatworms. Bilaterally symmetrical animals with a flat body; **dorsal** (top) and **ventral** (bottom) surfaces differ; no circulatory system needed because every part of body is near a surface. **Anterior** (front) end differs from **posterior** (hind) end. Sense organs and brain are concentrated at front end **(cephalization)** to form a **head**. A single all-purpose gastrovascular cavity, as in coelenterates; single opening functions as both mouth and anus. A simple, ladderlike nervous system, more concentrated at head end. Excretory tubules are **flame cells** (their beating cilia resemble a flickering flame). Three germ layers (Key 33): **ectoderm** (outer epidermis); **endoderm** (lining of gut); **mesoderm** (a loose **mesenchyme** in flatworms). **Acoelomate** (no body cavity, see Key 81). Many flatworms can regenerate missing parts following injury.
- Class **Turbellaria**. Mostly free-living; digestive tracts and sense organs still present; mouth often in middle of ventral surface.
- Class **Trematoda**. Small parasitic worms (flukes) with small, oval bodies; digestive tract simple, with mouth at anterior end.
- Class **Cestoda**. Tapeworms: highly degenerate internal parasites; greatly reduced digestive tract, nervous system, and sense organs.

Related phyla, also without body cavities:
- **Phylum Mesozoa:** Small, marine parasites with very few cells.
- **Phylum Rhynchocoela or Nemertea:** Worms with a long, sticklike proboscis or **evert** that can be withdrawn by turning inside out.
- **Phylum Gnathostomulida:** Small worms; outer epidermal cells each have a single cilium; mouth with paired, cuticle-hardened jaws.
- **Phylum Conodonta:** Small, extinct worms, known mostly from their teeth.

Key 81 Evolution of the coelom

OVERVIEW *Body cavities are useful in burrowing and as a hydrostatic skeleton. Various embryological origins are possible. Pseudocoels, as in roundworms and rotifers, arise from persistent blastocoel cavities. True coeloms are lined with mesoderm throughout and may arise either from out-pouching of the gut or from splitting within the solid mesoderm.*

Usefulness of fluid-filled body cavities (regardless of origin):
- In support, as a hydrostatic skeleton (Key 43);
- In burrowing, where inflation of the cavity can swell and anchor part of the body, or else wedge forward and push sediment aside.

Pseudocoel: A body cavity with both endoderm and mesoderm in its lining, derived from persistence of the blastocoel cavity (Key 33).

True coelom: A body cavity lined with mesoderm throughout, either an **enterocoel**, derived from outpouching of the gut (as in starfish), or a **schizocoel**, arising within the mesoderm by splitting (as in mammals). Animal phyla listed in Keys 82–93 all have a true coelom.

Pseudocoelomate phyla: Animals with a pseudocoel. Digestive tract (when not reduced as a parasitic adaptation) is usually a complete tube. This "assembly line" set-up permits regional specialization of function along the path from **mouth** to **anus**.

Phylum Aschelminthes: Small, wormlike animals with a pseudocoel; body usually covered with a chemically resistant **cuticle**.
- **Class Nematoda or Nemathelminthes:** Roundworms; the largest wormlike group. Free-living or parasitic; ends of body usually tapered.
- **Class Kinorhyncha:** Kinorhynchs.
- **Class Gastrotricha:** Gastrotrichs.
- **Class Gordiacea:** Horsehair worms.
- **Class Rotifera:** Rotifers, with a wheel-like crown of cilia at one end.

Phylum Acanthocephala: Spiny-headed parasitic worms.

Phylum Entoprocta: Small animals with a U-shaped digestive tract and a crown of tentacles **(lophophore)** surrounding both mouth and anus.

Key 82   Lophophorate phyla

OVERVIEW   *Phoronids, bryozoans, and brachiopods all have a true coelom and a ciliated feeding organ or lophophore. All are filter feeders that strain small particles of food from the water.*

Lophophore: A crown of ciliated tentacles that help gather suspended food particles. The cilia trap these particles and bring them to the mouth. Animal can withdraw lophophore if conditions are muddy or if predators threaten.

Similarities of phoronids, bryozoans, and brachiopods:
- All have lophophores.
- **True coelom** (Key 81), used as a hydrostatic skeleton (Key 43).
- Simple, U-shaped digestive tract, complete with mouth and anus.
- **Benthonic** (bottom-living), either burrowing or **sessile** (attached).

Phylum Phoronida (phoronid worms): Tube-dwelling worms with a lophophore surrounding the mouth. Coelom used in burrowing: muscular contraction builds up pressure in coelom, which swells sideways and pushes sediment aside. Probably related to ancestry of Bryozoa and Brachiopoda.

Phylum Bryozoa ("moss animals"): Largest and most successful lophophorate group; members quite varied. All are small, aquatic animals living in colonies. Many colonies are **polymorphic**, containing several dissimilar types of individuals. Mineralized exoskeletons (Key 43) are common. Ancestry uncertain, but probably close to Phoronida.

Phylum Brachiopoda: Probably derived from phoronid-like ancestors by addition of two-part shell as an aid in burrowing. Shell has two unequal valves; axis of symmetry bisects center of each valve. Valves are connected by muscles only (**class Inarticulata**) or by a hinge (**class Articulata**). Muscular stalk or **pedicle** attaches animal to the bottom. Fossil record shows these animals were more diverse and more numerous during the Paleozoic era (600 to 300 million years ago).

Key 83   Molluscs

OVERVIEW   *Molluscs are a large and diverse group of animals. Familiar molluscs include snails, clams, squids, and octopuses.*

Phylum Mollusca (molluscs): Animals with a true coelom of the schizocoel type (Key 81), usually bearing a shell composed mostly of calcium carbonate and secreted by a **mantle**. The mantle is always withdrawn at rear to form **mantle cavity** which contains anus and gills. Primitive molluscs and gastropods use a tongue-like **radula** with embedded teeth to scrape encrusted algae from rock surfaces.

Class Monoplacophora (primitive molluscs): Molluscs with a simple dome-shaped or low conical shell. Muscles, blood vessels, and other structures segmentally arranged. Digestive tract simple.

Class Gastropoda (snails and slugs): Body usually undergoes asymmetrical **torsion** (twisting or coiling). One-piece (univalve) shell, usually coiled. Most species herbivorous. Well-developed head, sense organs, and nervous system. Locomotion typically by creeping on a muscular **foot**.

Class Polyplacophora (chitons): Simple, flattened body, with shell divided into several overlapping plates that permit some flexibility. Head small but radula well-developed and used in feeding.

Class Bivalvia or Pelecypoda (clams and other bivalves): Body symmetrical, narrowly compressed from side to side. Two-piece (bivalve) shell; left and right valves are often mirror images (except at hinge). Many species filter-feed, straining small particles from water. Head and sense organs poorly developed. Muscular **foot** hatchet-shaped (flattened side-to-side), often used in burrowing.

Class Scaphopoda (tusk-shells): Small molluscs with tusk-like shells. Mantle cavity runs along entire length of shell on posterior margin; water passes through mantle cavity, exiting through hole at top.

Class Cephalopoda (octopuses, squids, and nautiloids, etc.): Body usually symmetrical. One-piece shell is symmetrically curved or coiled in median (midline) plane, or often lost. Most species are actively swimming predators. Head very well developed, with sense organs (especially eyes), brain, and beak. Muscular foot subdivided into numerous **tentacles**. Body doubled over, with mantle cavity (originally rear) tucked beneath head and opening forward. Frequent ''ink glands,'' secrete dark, inky fluid to confuse predators.

Key 84   Annelids and segmentation

OVERVIEW   *The Annelida are worms with segmentation of the body.*

Metamerism: Division of the body into numerous similar segments.

Phylum Annelida: Segmented worms. Complete digestive tract (with both mouth and anus) runs nearly the entire length of the body. Outer covering of **chitin** is thin, flexible, and prevents fluid loss. True coelom (Key 81) and most internal organs are segmentally arranged. Blood circulates in closed vessels only. Advanced excretory organs (nephridia) are present. Some ability to regenerate missing parts after injury.

Locomotion in annelids: Controlled separately in each segment:
- Each segment contains a walled-off portion of the body cavity.
- Muscles parallel to the body axis can shorten segments; these segments swell and anchor into the surrounding sand or soil.
- Muscles perpendicular to the body axis will lengthen certain segments and cause them to push forward.
- The nervous system produces rhythmic waves of shortening and waves of lengthening among the segments.
- Small bristles **(setae)** sometimes help anchor the shortened segments.

Class Polychaeta: Largest group, mostly marine; sense organs and nervous system highly developed, several setae per segment.

Class Oligochaeta (earthworms): Poorly developed head, only one pair of setae per segment. Important to soil because their digestive wastes leave behind soil nutrients and their tunnels let air reach plant roots.

Class Hirudinea (leeches): Mostly parasitic, live in fresh water, attach to the outside of animals and suck blood. Leeches have degenerate anatomy: fewer sense organs, fewer segments, etc.

Related minor phyla:
- **Phylum Priapulida:** Peanut-shaped worms.
- **Phylum Sipunculida:** Marine worms with tentacles around mouth.
- **Phylum Echiurida:** Cylindrical "proboscis worms."
- **Phylum Tardigrada:** "Water bears," with 8 short legs ending in claws.
- **Phylum Pentastomida:** Endoparasites inside vertebrates, with 2 pairs of short, degenerate legs armed with claws.

Key 85 Arthropods

OVERVIEW *Arthropods, the largest and most successful phylum of organisms, are characterized by a tough exoskeleton and jointed legs. The Onychophora are a transitional group between annelids and arthropods. Other major groups include the extinct trilobites, the largely aquatic Crustacea, the chelicerate groups (spiders, mites, scorpions, horseshoe crabs, etc.), the centipedes, the millipedes, and the insects. The insects alone make up about 75 percent of all the living species on Earth.*

Phylum Onychophora: A transitional group between annelids and arthropods. Segmented, wormlike body. Numerous short feet (one pair per segment), ending with claws. Feet around mouth function in seizing and tearing food.

Phylum Arthropoda: Animals with a tough **exoskeleton** (Key 43), often strengthened by calcium salts, and **jointed legs** with movable joints between rigid segments. Metamerism (Key 84) retained from annelid ancestors, but segments differ very much regionally. Mouthparts often derived from legs. Open circulatory system. Several anterior segments commonly coalesced into a **head**. Nervous system reminiscent of annelids, with ventral nerve cord, esophageal ring, and dorsal brain.

Trilobites: Extinct, marine arthropods with numerous similar **biramous** (two-branched) appendages, each with a leg-like part and a feathery gill. Body divided into **cephalon** (head), **thorax**, and **pygidium**. May be ancestral to other arthropods (but experts disagree on this). Flat-bottomed shape shows that many trilobites were bottom dwellers.

Crustacea: A largely marine group of arthropods, breathing by gills. Always two pairs of antennae. First post-oral segment has a pair of **mandibles**. Appendages **biramous** (two-branched), as in trilobites. Includes lobsters, crabs, shrimp, barnacles, and many other species.

Chelicerate groups: Originally marine, but one group successfully invaded land environments. No antennae are present. First 2 pairs of appendages include a pair of **chelicerae** (piercing structures that may be used to inject venom) and a pair of **pedipalps** which hold prey while the chelicerae pierce them. Many chelicerates are predators.

Body usually divided into cephalothorax and abdomen (except in mites). Includes **Pycnogonida** (sea spiders), **Xiphosura** (horseshoe crabs), and **Arachnida** (scorpions, spiders, mites, etc.).

Uniramous "myriapods": Arthropods with elongate, wormlike bodies and many pairs of legs. Includes centipedes **(Chilopoda)**, millipedes **(Diplopoda)**, and two other groups **(Pauropoda, Symphyla)** allied to insects. All myriapods have similarities that they share with insects: a single pair of antennae; a pair of mandibles; **uniramous** appendages (having only a single branch); a type of air-breathing using a tracheal system of air-tubes (Key 38); usually terrestrial habitats, often hiding beneath rocks and rotting logs.

Insects: The largest and most successful group of arthropods, including about 75% of all species in the entire animal kingdom. Mandibles, uniramous appendages, and tracheal systems as in myriapods. Division of the body into three portions; **head** (containing antennae, mandibles, and compound eyes); **thorax** (typically with three pairs of walking legs and two pairs of wings); and an **abdomen** of seven or more segments (containing reproductive structures).

Key 86   Echinoderms

OVERVIEW   *Echinoderms (starfishes, crinoids, sea urchins, and their relatives) are often radially symmetrical as adults, but their embryonic stages show similarities to the chordates.*

Phylum Echinodermata: Animals with a unique **water-vascular system**, using sea water as a circulatory fluid. Several embryonic similarities to chordates (Key 87), including a true coelom which develops as an enterocoel (Key 81). Change of symmetry in many cases, from a bilateral larva to a radial adult, typically in a five-fold pattern. Protective plates or shells frequently made of calcium carbonate and armed with bumps or spines. High ability to regenerate lost parts.

Sessile (attached) echinoderms (Homalozoa and Crinozoa): Echinoderms that grow attached include **crinoids** (sea lilies) and a variety of extinct groups (blastoids, cystoids, carpoids, etc.). Many grow on stalks attached to the bottom. Body cup-shaped, open toward the top, with a mouth in the center of the top surface. Arm-like **rays**, in multiples of five, grow out and upward from the margins of the mouth. Each ray has a ciliated groove that traps small food particles and brings them to the mouth. The earliest fossil forms were irregular and lacked symmetry, but radial symmetry developed early, generally in a five-fold pattern. Biologists believe that echinoderm ancestors were bilaterally symmetrical and that **filter feeding** (filtering small particles of food from the water) made radial symmetry selectively advantageous. Attached echinoderms flourished mostly during Paleozoic times. Today, only a few crinoids remain; other attached echinoderms are extinct.

Free-moving echinoderms (Echinozoa and Asterozoa): Mostly bottom-feeding scavengers and predators that attack other invertebrates. The mouth, on the lower surface, faces downward. Branches of the water-vascular system may form foot-like **podia**, used in locomotion.
- **Asterozoa:** Body star-shaped, with protruding arms. Includes starfishes and brittle stars.
- **Echinozoa:** Body globe-shaped, with no protruding arms. Includes sea cucumbers, sea urchins, and sand dollars.

Related minor phyla:
- **Phylum Chaetognatha:** Arrow worms, with a dart-shaped body.
- **Phylum Pogonophora:** Deep-water worms living in chitinous tubes.

Key 87   Lower chordates

OVERVIEW   *The Chordata include animals with a noto-chord, a dorsal hollow nerve cord, gill slits, and many embryological similarities linking them with echinoderms. Chordates include acorn worms, tunicates, sea lancets, fishes, amphibians, reptiles, birds, and mammals.*

Notochord: A stiff, flexible rod, forming the body axis. When muscles contract, it prevents the body from collapsing like an accordion. In embryos, it induces nervous system to form above it (Key 34).

Gill slits: Openings from **pharynx** to either side, just behind mouth.

Hemichordata: Acorn worms and their relatives. All of them filter feed. Some use gill-slits; others use tentacle-like feeding structures. Related to Chordata, but now usually treated as a separate phylum.

Phylum Chordata: Animals with a notochord, a series of gill slits, and a dorsal, hollow nerve cord developing from a **neural tube**. These traits may occur in larval stages, not always in adults.

Deuterostome characteristics: Embryological similarities shared by chordates, hemichordates, and echinoderms:
- **Radial cleavage:** the eight-celled stage has two tiers of four cells each, with each cell directly above or below another.
- **Indeterminate cleavage:** cells separated in early embryonic stages can develop into an entire embryo.
- **Deuterostome condition:** the embryo's blastopore (Key 33) becomes the posterior (tail) end. (In molluscs, arthropods, and other **protostome** phyla, the blastopore becomes the mouth.)
- Many primitive members of both phyla are sessile filter-feeders.

Urochordata (tunicates or "sea squirts"): An actively swimming larva with well-developed notochord and nerve chord undergoes metamorphosis into a filter-feeding adult. The adult usually passes large amounts of water through a large gill basket.

Cephalochordata (sea lancets or amphioxus): Small, thin animals that filter feed by passing water through many gill slits. A notochord extends the entire length of the animal, including the head.

Vertebrata (vertebrates): Animals with a **vertebral column** or **backbone** that functionally replaces the notochord in adults and a **braincase** that encloses and protects the brain. *Examples:* fishes, amphibians, reptiles, birds, and mammals (Keys 88–93).

Key 88 Fishes

OVERVIEW *Fishes are aquatic vertebrates that have fins and gills throughout their adult lives. They are a very heterogeneous assemblage of four distinct classes: the jawless fishes (Agnatha); the extinct, armored Placodermi; the shark-like or cartilaginous fishes (Chondrichthyes); and the bony fishes (Osteichthyes).*

General characteristics of fishes:
- Central nervous system protected by being enclosed inside skeleton.
- Free-swimming by means of side-to-side undulations of the body.
- Gills used in respiration; most fishes no longer filter-feed.

Class Agnatha: Jawless fishes, often with a filter-feeding larval stage. Extinct forms (''ostracoderms'') were often heavily armored and continued to filter-feed as adults. Modern forms **(cyclostomes)** include lampreys and hagfishes. Adults are eel-shaped parasites of other fish: lampreys suck blood; hagfishes eat their way through the flesh of their victims.

Class Placodermi: An extinct group in which jaws first evolved. Paired fins also evolved in this group and are retained in all further vertebrate classes. Many placoderms were predators, from 6 inches up to 50 feet long.

Class Chondrichthyes: Cartilaginous fishes, including sharks, skates, and rays. Bone is reduced to a series of toothlike denticles embedded in the skin. The rest of the skeleton is made of cartilage only.

Class Osteichthyes: Bony fishes, including the vast majority of fishes. Scales and internal skeleton are both usually bony. A wide diversity of sizes and shapes occurs in this group.

Key 89 Amphibians and reptiles

OVERVIEW *Amphibians have aquatic, gill-breathing larvae and lung-breathing adults; their eggs are laid in the water. Reptiles all have a type of shell-covered egg that must be laid on land. Reptiles were the dominant land animals of the Mesozoic Era; their descendants include the birds and mammals.*

Class Amphibia: Eggs are laid in contact with fresh water, then fertilized externally. Larvae (''tadpoles'') breathe with gills, then undergo metamorphosis into an adult, usually with lungs and legs. Living species always have slippery, moist skin. *Examples:* salamanders, newts, frogs, toads, and extinct labyrinthodonts.

Origin of amphibians: Amphibians evolved from a group of bony fishes (Osteichthyes) called crossopterygians who already had lungs and internal nostrils. The critical change transformed the fleshy fins into walking legs.

Class Reptilia: Reptiles differ sharply from amphibians in laying a shell-covered type of egg (**amniote egg** or **cleidoic egg**). These eggs must be laid on land, or else hatched inside the female's body. The shell, secreted by the female, prevents the passage of liquids. Since the sperm must swim in a liquid medium, it must enter the egg first; this requires **internal fertilization**. Reptiles also have a dry, tough (leathery or scaly) skin. Legs are generally stronger than in amphibians; most toes have claws. Body temperature changes to match surroundings. *Examples:* turtles, snakes, lizards, dinosaurs, and crocodiles.

Embryonic membranes: Reptiles, birds, and mammals all have amniote eggs within which several specialized membranes develop from the embryo:
- **Chorion:** Provides protection.
- **Amnion:** Encloses the embryo in its own protective watery bag; gives the amniote egg its name.
- **Allantois:** Serves as a lung-like respiratory organ for the embryo.

Key 90   Birds

OVERVIEW   *Birds are vertebrates adapted for flight. Important adaptations include feathers, wings, warm-blooded metabolism, hollow bones, and loss of teeth and tail bones.*

Class Aves (birds): Warm-blooded, feathered vertebrates.

Flight adaptations: Most birds have adaptations for flying, including wings, feathers (modified scales), and good vision. Modern flying birds also have strong ribs and a rigid sternum (breastbone) with a keel.

Metabolism: A high rate of metabolism is needed for flight. Birds are warm-blooded, meaning that their metabolic rates and body temperatures are always rather high, regardless of external temperature. Downy feathers are part of an insulating layer that makes a high internal body temperature possible. The complete separation of oxygen-rich and oxygen-poor blood also increases metabolic efficiency.

Weight reduction: Modern birds have various adaptations that reduce weight, including reduction of the tail bones, loss of the teeth and lightening of the jaws, development of hollow air spaces in the arm bones, and loss of one ovary in female birds.

***Archaeopteryx* and the origin of flight:** *Archaeopteryx,* the oldest bird, was preserved in a fine-grained Jurassic limestone. It had many reptile characteristics, including a long tail, slender ribs, weak sternum, small braincase, and jaws with teeth. However, it also had feathers and was probably capable of flight, and is, therefore, classified as a bird.

Modern birds:
- Modern flightless birds include the ostriches, rheas, moas, and kiwis.
- Modern flying birds include owls, storks, gulls, pigeons, eagles, hawks, woodpeckers, and the many perching songbirds (Passeres), which make up the largest group.

Key 91 Mammals

OVERVIEW *Mammals are warm-blooded vertebrates that nurse their young with milk. The right and left sides of the four-chambered heart are completely separate. Mammals originated from reptiles.*

Class Mammalia: Vertebrates covered with insulation, usually hair or fur (occasionally blubber). Metabolic rate and body temperature are kept high (**homeothermy**). Glands in skin secrete sweat and oily secretions (sebum). Young mammals are nursed by their mothers; **milk** is secreted by **mammary glands**, derived from sweat glands. Normal standing posture keeps the body elevated from the ground, compared to the low-slung posture of amphibians and reptiles. Three ossicles, or tiny ear bones (malleus, incus, stapes), found in the middle ear. Four-chambered heart has complete separation of oxygen-rich and oxygen-poor blood. Only one bone, the dentary, makes up the lower jaw (mandible) on each side. A muscular diaphragm is responsible for most breathing movements. A bony hard palate separates nasal cavity from oral cavity, permitting breathing and chewing at the same time. Teeth vary in shape with their position in the mouth and are restricted to only two waves of growth and replacement instead of many. Brain and braincase are both larger than in reptiles.

Origin of mammals: Mammals originated from mammal-like (synapsid) reptiles, probably during the Triassic period. The transition involved changes in the teeth and tooth replacement, the replacement of one jaw hinge (between articular and quadrate bones) with another (between squamosal and dentary bones), and the conversion of the articular and quadrate bones into the malleus and incus.

Monotremata: Egg-laying mammals. *Example:* Platypus.

Marsupials: Pouched mammals. *Examples:* Kangaroo, opossum.

Placental mammals (Eutheria): Mammals in which the fetus is nourished *in utero* by a placenta. This group includes the vast majority of mammals, arranged in over 30 orders, about half extinct and half with living members. *Examples:* Shrews, mice, rats, bats, cows, deer, pigs, dogs, cats, monkeys, humans, whales, horses, elephants, rabbits.

Key 92   Primates

OVERVIEW   *Most primate characteristics arose as adaptations to life in trees. These features include grasping hands and feet, opposable thumbs, reliance on vision, expansion of the brain, higher intelligence, increasing emphasis on learned behavior, single births, and greatly increased parental care. The group to which we belong, the catarrhines, have protruding noses, reduced tails, and only two premolars in each jaw. Catarrhines also include apes and Old World monkeys, both confined to the Eastern Hemisphere.*

Order Primates: Monkeys, apes, humans, lemurs, tarsiers, and related animals.

Primate characteristics (mostly related to **arboreal adaptations**, adaptations to life in the trees):
- Arboreal locomotion
- Grasping hands and feet (which wrap around branches)
- Opposable thumb and/or big toe (wrap around in opposite direction from other digits)
- Increased freedom of rotation in forearm
- Increased reliance on vision, and less on smell
- Binocular, stereoscoptic vision (in depth)
- Expanded visual centers in brain; more folds in brain surface
- Visual inspection and manipulation of objects
- Increased intelligence
- Increased emphasis on learned behavior; juvenile inexperience
- Increased length and intensity of parental care
- Uteri (Key 41) fuse to a single "uterus simplex"

Plesiadapoidea or Paromomyiformes: Archaic primates.

Lemuroidea or Strepsirhini: Lemurs, lorises, and galagos.

Tarsioidea: *Tarsius* and its extinct relatives.

Platyrrhini: New World monkeys and marmosets; three premolars in each jaw, flat noses, and strong tails that aid in locomotion.

Catarrhini: Old world monkeys, apes (gibbons, orangutan, gorilla, chimpanzee), and humans, with two premolars in each jaw, protruding noses (nostrils opening downward), and reduced tails, living in Africa, Asia, and Europe.

Key 93   Human evolution

OVERVIEW   *Humans walk upright. The many consequences of this include tool use, speech, and anatomical changes such as a gently S-shaped spinal column (with a lumbar curve) and a more rounded cranium from which the spinal cord exits at the bottom.*

Family Hominidae (humans): Catarrhine primates distinguished from apes (family Pongidae) principally by the attainment of upright posture. Characteristics include: upright, bipedal locomotion (walking, running); larger and more rounded braincase; spinal column exits (through foramen magnum) at the bottom, not the rear, of the skull; reduced spines on neck vertebrae; spinal column gently S-shaped, with a lumbar curve (concave to the rear along the lower back); pelvis wider; iliac crests expanded; gluteus maximus muscle enlarged and rotated to the rear (pulls leg to rear instead of to side); canine teeth reduced (tools are now major weapons); lower jaw symphysis strengthened by a chin; tooth rows rounded instead of parallel; habitual use of tools (hands are free to hold them); habitual use of language.

Origin of the Hominidae: Occurred approximately between 4 and 5 million years ago, when upright posture was attained. Human footprints at Laetoli, Kenya, are about 4.1 million years old.

Australopithecus: The earliest hominids, known from Africa. Most paleontologists now consider *Australopithecus* to be an extinct side branch, not an ancestor of *Homo.* One of the oldest specimens, nicknamed "Lucy," was only about 4 feet tall but walked upright. Other specimens are known from both South Africa and East Africa.

Homo habilis: An East African contemporary of *Australopithecus,* from about 4.5 to about 1.5 million years ago. Body size small, about 4 feet tall. Perhaps responsible for the earliest stone tools.

Homo erectus: Lived in the middle Pleistocene, after the extinction of *Australopithecus.* Fossils known from Africa, Europe, and Asia. In a cave near Beijing, China, heat-fractured rocks show that fire was used.

Homo sapiens: First appeared in the late part of the Ice Age. Larger skull size than earlier species. Used more advanced tools. Invented agriculture around 6,000 years ago in several places.

Glossary

*Included here are the definitions of many, but not all, of the terms used in the keys. Terms printed in **boldface** in the definitions are further explained under their own entries in this glossary. For terms not listed here, please consult the index.*

Accessory pigments
Energy-absorbing plant pigments other than chlorophyll.

Acetyl coenzyme A
A molecule that combines with **pyruvate,** then enters the Krebs cycle.

Acoelomate
Lacking a body cavity.

Acquired characteristics
Changes occurring during an organism's lifetime.

Active site
Part of an **enzyme** that fits together with a **substrate**.

Active transport
Transport *against* a **concentration gradient**, requiring energy.

Adaptations
Features that help organisms cope with and exploit their environments.

Adaptive radiation
Diversity among the descendants of a single species.

Adenosine triphosphate (ATP)
An important energy-rich molecule in biological systems.

Aerobic
Conditions in which oxygen is present.

Algae
Aquatic **eucaryotic** organisms with plastids but no **embryos**.

Allele
One of the several possible variants of a gene.

Alternation of generations	A life cycle in which **gametophyte** and **sporophyte** stages follow one another alternately.
Amino acid	An organic acid containing an amino group ($-NH_2$), and serving as a building block for **proteins**.
Amoeboid locomotion	Locomotion using **pseudopods**.
Anaerobic	Conditions in which oxygen is not present.
Analogy	Similarities resulting from similar adaptations.
Aneuploidy	Loss or gain of one chromosome at a time (compare **polyploidy**).
Angiosperm	A plant reproducing by means of **flowers**, with **seeds** enclosed in **ovaries**.
Anterior	The front end of a bilateral animal.
Antibody	An infection-fighting protein released in response to an **antigen**.
Anticodon	A three-nucleotide sequence of tRNA that pairs with a **codon**.
Antigen	Anything capable of evoking an immune response.
Arboreal	Living in trees.
Archenteron	A cavity in the gastrula which later develops into the gut.
Artificial selection	Selection of domesticated species by humans.

Atom	Smallest particle of an element.
ATP (adenosine triphosphate)	An important energy-rich molecule in biological systems.
Autosomal	Not sex-linked.
Autotroph	Organism that can derive energy from sunlight and carbon compounds from CO_2.
Balanced polymorphism	Persistence of several alleles in a population when heterozygotes are selectively favored.
Bilateral symmetry	Symmetry in which right and left halves are mirror images.
Binomial nomenclature	Giving each species a two-word name.
Biome	Group of ecologically similar communities (e.g., tundras or deserts) with ecologically similar species.
Biosphere	The largest ecosystem, consisting of the Earth and all its inhabitants.
Biramous	Two-branched.
Blastopore	Entrance to the **archenteron** of an embryo.
Blastula	Embryonic stage consisting of a hollow ball of cells.
Blood	A liquid connective tissue containing **erythrocytes, leucocytes**, and platelets.
C_3 plants	Plants that use the **Calvin cycle**, incorporating CO_2 into a 3-carbon compound.

C₄ plants

Plants like sugarcane that incorporate CO_2 into a 4-carbon compound.

Calvin cycle

A series of reactions in which CO_2 is incorporated into a 3-carbon compound.

Cambium

Persistently embryonic tissue (**meristem**) growing between the **xylem** and **phloem** in many plants.

Carbohydrates

Sugars, starches, and related compounds.

Carcinogen

Cancer-causing agent.

Carnivores

Animals that eat other animals.

Carriers

In genetics, **heterozygous** individuals who pass a gene to their offspring without showing any of its effects themselves.

Catalyst

Something that speeds up a chemical reaction, but is not itself used up or altered as a result.

Cell

The smallest unit of a living thing, bounded by a cell membrane, containing DNA and RNA, and carrying on all major life functions.

Cell cycle

The cycle which includes both cell division (**mitosis** and **cytokinesis**) and **interphase**.

Cell membrane (=plasma membrane)

The outer lining of each cell.

Centriole

A ''9+2'' arrangement of microtubules from which spindle fibers form during **mitosis**.

Centromere	The central part of a **chromosome**, by which the **chromatids** are held together.
Chemical evolution	The theory of life's gradual origin from organic chemicals formed in a reducing (hydrogen-rich) atmosphere.
Chitin	A protein-polysaccharide complex found in exoskeletons.
Chloroplast	Photosynthetic plant organelle containing chlorophyll.
Chordata	A phylum including **vertebrates** and their relatives.
Chromatid	One of the strands of a **chromosome**.
Chromatin	DNA-rich material within the nucleus, organized into **chromosomes**.
Chromoplast	Organelle containing plant pigments other than chlorophyll.
Chromosomes	Elongated structures containing hereditary material (**DNA**) in the form of **genes**.
Cilia	Hairlike organelles on the surfaces of certain cells.
Cline	A geographical character gradient (running north to south, for example).
Codon	Part of an mRNA message, composed of three nucleotides.
Coelom	A fluid-filled body cavity lined entirely with **mesoderm**.
Cofactor	The nonprotein component of certain enzymes.

Cold-blooded

Having a body temperature which fluctuates close to that of the external environment.

Colinearity

Correspondence in the sequences of DNA, mRNA, and protein.

Community

All species that live together and interact in a particular habitat.

Complement

A group of proteins that activate phagocytosis, inflammation, and cell lysis.

Compositionist

One who insists that the study of parts alone will never reveal the behavior of the whole systems made of them.

Concentration gradient

A difference in concentration.

Connective tissue

Type of tissue containing large amounts of extracellular material (**matrix**).

Convergence

Similar adaptational results in unrelated **lineages**.

Covalent bond

Chemical bond formed by a pair of shared electrons.

Crossing over

The breakage, switching, and rejoining of the arms of **homologous chromosomes** during prophase of meiosis I.

Cuticle

A chemically resistant, waterproof outer coating.

Cytochromes

Electron acceptors used in the electron transport system.

Cytokinesis

Division of the cytoplasm, following **mitosis**.

Cytoplasm

All of the cell outside the nucleus.

Cytoskeleton

The structural framework of a cell, including **microtubules** and **microfilaments**.

Deuterostome

Condition in which the embryonic **blastopore** becomes the hind end and the mouth develops as a secondary structure at the opposite end.

Diffusion

Transport (usually passive) down a **concentration gradient**.

Diploid (2N)

Having two chromosomes of each type.

DNA (deoxyribonucleic acid)

A nucleic acid containing deoxyribose sugar.

Dominant

An allele expressed in the phenotype even when only one such allele is present, or the phenotype so formed.

Duplication

Mutation in which part of a chromosome occurs twice.

Ecosystem

A **community** plus its physical surroundings.

Ectoderm

Outer layer of cells in an embryo or a coelenterate; its derivatives include epidermis of the skin, nervous system, etc.

Egg

A female gamete (**ovum**).

Electrophoresis

A technique that separates proteins according to their speed of migration through an electrical field.

Embryo

The early development stages of an organism, before organs are formed; in plants, a reproductive cell surrounded by nonreproductive cells.

Endochondral ossification Bone formation within cartilage.

Endocrine
Secreting a product (called a **hormone**) into the blood stream.

Endoderm
Inner layer of cells in an embryo or a coelenterate; its derivatives include inner lining of the gut, lungs, liver, etc.

Endoplasmic reticulum
A series of folded membranes functioning in the internal transport of materials within a cell.

Endosperm
Triploid (3N) tissue in a **seed**, containing stored food.

Endosymbiosis
Theory of the origin of **eucaryotic** organelles from **procaryotic** organisms.

Enterocoel
A coelom that arises from the gut.

Enzyme
An organic **catalyst** made largely or entirely of protein.

Epidermis
An outer, protective layer of cells arranged as a flat surface.

Epistasis
A type of gene-gene interaction in which one gene "masks" or suppresses the phenotypic expression of a totally different gene.

Epithelium
Tissue which originates in broad, flat surfaces.

Erythrocytes
Red blood cells, containing **hemoglobin**.

Eucaryotic
Having cells with a well-defined nucleus (bounded by a nuclear envelope) and organelles (such as endoplasmic reticulum) composed of internal membranes.

Exobiology
The search for life outside planet Earth.

Exocrine   Secreting a product along a surface or into a duct.

Exon   An mRNA sequence after excision has taken place.

Exoskeleton   A skeleton on the outside, as in insects.

Extinction   Termination of a species or lineage without issue.

Falsifiable   Capable of being proven false by observation or experiment.

Fatty acids   Long-chain molecules containing $-COOH$ at one end.

Fauna   All the animals of a habitat or region.

Fermentation   Anaerobic metabolism giving off CO_2.

Fertilization   The union of egg and sperm cells to produce a **zygote**.

Fetus   A developing mammal after its organs are formed.

Filter feeding   Feeding by straining small, suspended food particles from the water.

Flagella   Whiplike or propeller-like organelle on the surface of certain cells.

Flora   All the plants of a habitat or region.

Flower   A plant **ovary** (enclosing one or more **seeds**) and its surrounding structures.

Fossil   Remains or other evidence of life of past geologic ages.

Fruit   A ripened plant **ovary** (or several together).

Gametes Egg or sperm cells.

Gametophyte A **haploid** body, or the haploid portion of any life cycle.

Gastrovascular cavity Single, all-purpose cavity in Cnidaria and other simple animals, with only one opening (mouth but no anus).

Gastrula Embryonic stage during which germ layers (**endoderm, ectoderm, mesoderm**) form.

Gene Part of a DNA molecule that codes for one protein and thus affects at least one trait.

Genetic drift Random changes in gene frequencies due to chance, especially in smaller populations.

Genotype All the genetic traits of an organism, as revealed by breeding experiments.

Genus A group of related species.

Geographic speciation Species formation in which reproductive isolation originates while the future species are geographically separated by a barrier to dispersal.

Gill slits Openings in the side of the pharynx in the phylum Chordata.

Glycolysis The breakdown of glucose into **pyruvate**.

Golgi apparatus A group of stacked, flattened vesicles that help package proteins into **vacuoles**.

Gonad An organ producing eggs or sperm, such as an ovary or testis.

Greenhouse effect	Warming of the Earth by the trapping of heat rays that enter the atmosphere and cannot escape.
Gymnosperms	Vascular plants with naked seeds.
Habitat	The place where a population lives.
Haploid (N)	Having only one chromosome of each type.
Hemocoel	The general circulatory cavity of an open circulatory system.
Hemoglobin	An oxygen-carrying pigment in red blood cells.
Herbivores	Animals that eat plants.
Hermaphrodite	An individual possessing both male and female body parts.
Heterotrophs	Organisms that need to feed on other organisms.
Heterozygous or heterozygote	Genotype with two unlike **alleles** of a particular gene.
Hill reaction	$H_2O + NADP^+ + energy \longrightarrow$ $O_2 + NADPH + H^+$
Histone	A type of protein contained in **eucaryotic** chromosomes.
Homeostasis	Tendency of living systems to maintain a balance that keeps themselves going.
Homeothermy	Warm-blooded condition in which a high internal body temperature is maintained.
Homologous pairs (homologous chromosomes)	Chromosomes that pair with one another in prophase I of meiosis.

Homology	Deep-seated resemblance reflecting common ancestry.
Homozygous or homozygote	Genotype with two identical **alleles** of the same gene.
Hormone	Chemical messenger secreted directly into the blood stream.
Hydrocarbon	Organic compound containing hydrogen and carbon only.
Hydrogen bonds	Weak bonds between a **polar** part of a molecule and the hydrogen atom of another molecule.
Hydrolysis	Process of splitting a molecule apart by inserting a water molecule.
Hypothesis	A statement subject to testing.
Immunoglobin	An infection-fighting protein (**antibody**).
Inbreeding	Increased mating among relatives.
Independent assortment	A principle in genetics that genes on different chromosomes **segregate** independently of one another.
Instinct	A complex, innate (inborn) behavior pattern.
Interphase	The interval between one mitosis and the next.
Intron	Part of an mRNA sequence that is removed during excision.
Invertebrates	Animals without a backbone.
K-selection	Selection operating on populations living near their carrying capacities (K).

Leucocytes	White blood cells.
Leucoplast	Colorless food-storage plastid in plants.
Lichen	A symbiotic association of an alga and a fungus.
Light reactions	Those reactions of **photosynthesis** that require light energy, principally the splitting of water and the release of oxygen.
Lineage	An ancestor-to-descendant sequence of species.
Linkage	The inheritance of two or more **genes** as a unit when they are located on the same **chromosome**.
Linkage group	All the genes located on the same chromosome.
Lipids	Fats and other organic compounds soluble in ether.
Locus	The location of a gene on a chromosome.
Lysogenic cycle	A cycle in which a virus replicates as part of the host cell's genetic material.
Lysosome	An organelle containing protein-digesting enzymes and surrounded by a membrane.
Lytic cycle	A cycle in which a virus replicates and releases thousands of copies of itself by rupturing the host cell.
Macroevolution	Evolution above the species level.
Mantle	In molluscs, a layer of cells that secretes the shell.

Matrix	Extracellular material in connective tissues.
Mechanist	One who views life as no more than a physical-chemical process.
Meiosis	Cell division in which the chromosome number is reduced in half.
Meristem	Embryonic plant tissue that continues growing throughout life.
Mesoderm	Middle germ layer, which forms muscles, skeleton, circulatory system, excretory system, reproductive system, etc.
Messenger RNA (mRNA)	A long strand of RNA synthesized in the nucleus and used in the cytoplasm for protein synthesis.
Metabolism	Process of deriving energy from the breakdown of energy-rich materials.
Metamerism	Division of the body into numerous similar segments.
Metamorphosis	A drastic developmental change from a larval stage to an adult.
Microevolution	Evolution below the species level.
Microfilaments	Contractile filaments of proteins like actin and myosin.
Microtubules	Hollow structures made of protein, important in cell motility.
Milk	Nutritive fluid fed by mammals to their young.
Mimicry	One species' false resemblance to another.

Mitochondria	Energy-producing organelles containing a folded inner membrane surrounded by an outer membrane.
Mitosis	Division of the cell nucleus which leaves the chromosome number unchanged.
Molecule	Smallest particle of a compound.
Molting	Shedding and replacing the outer protective covering.
Mosaic evolution	Evolution of different characters at different rates and times.
Multiple alleles	More than two **alleles** of a particular gene, as in the A-B-O blood group system.
Mutagen	A substance that causes mutations.
Mutations	Permanent changes that occur in hereditary material.
Natural selection	Inherited differences in the ability to survive and reproduce.
Negative control	A gene is turned ''off'' when needed, but is normally ''on.''
Neuron	A nerve cell.
Neurotransmitters	Chemicals released by the ends of neurons.
Neurulation	Formation of the nervous system.
Niche	The ecological role or ''way of life'' of a population.

Notochord

In the phylum **Chordata**, cells on the roof of the **archenteron** that act as an organizer for the nervous system and that later form a stiff but flexible rod along the body axis.

Nuclease

An enzyme that causes breaks in DNA sequences.

Nucleolus

RNA-rich region within the nucleus in which ribosomes are formed.

Nucleotide

A building block of **DNA** or **RNA**, composed of phosphate, a sugar, and a nitrogen-containing base.

Nucleus

The part of a cell containing hereditary material in the form of genes.

Operon

A gene, or group of related genes, regulated together as a unit.

Organelle

A cytoplasmic structure within a cell.

Organic compound

A compound containing covalently bonded carbon.

Organizer

A substance secreted by one group of cells which induces growth changes in other cells.

Ovary

A female organ producing gametes (ova), or (in plants) a protective structure enclosing a seed.

Ovum

A female gamete.

Oxidative phosphorylation

The synthesis of **ATP** from ADP.

Paleontology   The study of fossils.

Parallelism   Independent occurrence of the same or similar **trends** in different **lineages**.

Passive transport   Transport without the use of energy from a region of higher concentration to a region of lower concentration.

Peptide bond   A bond joining adjacent **amino acids**, as in a **protein**.

Perikaryon   Part of a nerve cell body surrounding the nucleus.

Phagocytosis   A type of bulk transport in which folds of cell membrane engulf food material, forming a vacuole.

Pharynx   Part of the digestive tract containing gill slits in the **Chordata**.

Phenotype   All the visible traits of an organism that can be revealed by examining it closely without any breeding.

Pheromone   A chemical used in communication, such as a sexual attractant.

Phloem   Tissue in **vascular plants** that carries photosynthetic products downward.

Photosynthesis   A process by which plants (and some bacteria) make sugars using sunlight.

Phylogeny   A family tree.

Phylum   A major subdivision of a kingdom.

Plasma membrane (=cell membrane)	The outer lining of each cell.
Plasmid	A piece of **procaryotic** DNA that can detach from the main chromosome.
Plastid	Plant organelles such as **chloroplasts, chromoplasts,** or **leucoplasts**.
Pleiotropy	A condition in which one gene affects many phenotypic traits.
Polar body	One of the small cells produced, together with the ovum, during meiosis in females.
Polar	Having an uneven distribution of electrical charges (one end more negative than the other).
Pollen	Structures enclosing male **gametophytes** in higher plants.
Polypeptide	A long chain of **amino acids** linked by **peptide bonds**.
Polyploidy	Addition of entire haploid sets of chromosomes.
Population	All those members of a species that can breed with one another.
Positive control	A gene is turned "on" when needed, but is normally "off."
Posterior	The hind end of a bilateral animal.
Predation	Interaction between species that benefits one at the expense of the other.
Primary consumers (herbivores)	Animals that eat plants.

Primary producer	Organism that can use sunlight to incorporate CO_2 into organic compounds.
Procaryotic	Having cells lacking any well-defined, membrane-bounded nucleus, any endoplasmic reticulum, or other internal membranes.
Protein	An organic molecule built of **polypeptides** (chains of **amino acids**).
Protostome	Condition in which the embryonic **blastopore** becomes the mouth.
Pseudocoel	A body cavity lined with both **meosderm** and **endoderm**.
Pseudopods	Cellular extensions used in locomotion.
Punctuated equilibrium	A theory of rapid evolutionary change alternating with steady equilibria.
Pyruvate (pyruvic acid)	A compound, CH_3—CO—$COOH$, into which glucose is initially broken down.
r	The ''intrinsic rate of natural increase'' in the exponent of the growth equation.
r-selection	Selection by frequent density-independent mortality, favoring prolific, rapid reproduction and high values of r.
Radial symmetry	Pattern with many planes of symmetry arranged about an axis.
Recessive	An allele expressed in the phenotype only when two such alleles are present, or the phenotype that they form.
Reductionist	One who explains whole structures in terms of their parts.

Regions	Continental areas inhabited by related species.
Regulation	Controlling whether or not a gene produces a protein sequence.
Replication	The copying of **DNA** to make more DNA.
Reproductive isolating mechanisms	Any characteristics of a species that prevent its interbreeding with other species.
Restriction endonuclease	A **nuclease** enzyme that attacks only certain DNA sequences.
Ribosomal RNA (rRNA)	RNA in ribosomes.
Ribosomes	Small organelles that function in protein synthesis.
RNA (ribonucleic acid)	A nucleic acid containing ribose sugar.
Root	Absorptive plant part which typically grows downward.
Saprobes (saprophytes)	Decomposer organisms that feed on dead or decaying organic matter.
Saprophytic	Living on dead or decaying matter.
Schizocoel	A coelom that arises from a split within the mesoderm.
Secondary consumers (carnivores)	Animals that eat other animals.
Seed	An easily dispersed embryonic **sporophyte** of a higher plant.
Segregation	In genetics, the separation of dominant and recessive alleles in the offspring of heterozygotes.

Selection	Any situation in which genotypes contribute unequally to the next generation.
Semiconservative	Half newly synthesized and half pre-existing, like a DNA molecule.
Sessile	Attached to the bottom.
Sex-linked	Carried on the X-chromosome.
Sexual selection	Selection based on success in mating.
Somatic cells	All cells except sex cells (**gametes**).
Speciation	Formation of new species by the splitting of an ancestral species.
Species	Groups of interbreeding natural populations that are reproductively isolated from other species.
Sperm	A male gamete.
Spermatogenesis	Meiosis in males, resulting in sperm cells.
Spontaneous generation	The sudden origin of life from nonlife.
Sporophyte	A **diploid** body, or the diploid portion of any life cycle.
Steroids	Chemicals derived from cholesterol, such as many sex hormones.
Stomate	Opening in the lower epidermis of a leaf, allowing air exchange.
Substrate	A chemical substance upon which an **enzyme** acts.
Supporting tissues	Tissues that support a plant and define its shape.

Sweepstakes dispersal	Dispersal of species across barriers by infrequent, low-probability (high-risk) events.
Symbiosis	An association of two species living together.
Synapse	The meeting of two neurons.
Syncytium	A structure containing many nuclei but no internal cell boundaries.
Taxes	Oriented locomotor movements.
Taxon	Any kingdom, phylum, class, order, family, genus, or species.
Taxonomy	The theoretical study of classification.
Tetrad	A four-stranded structure, consisting of two double-stranded **homologous chromosomes**, which forms in the first prophase of **meiosis**.
Tissue	A group of similar cells and their products, closely related in function and location.
Transcription	The copying of **DNA** information to **RNA**.
Transfer RNA (tRNA)	Short, twisted strands of RNA that usher amino acids into growing polypeptide sequences during protein synthesis.
Translation	The use of an **RNA** message to control the synthesis of a protein sequence.
Translocation	Joining of a chromosome fragment to a different chromosome.
Trend	Continued change of morphology within a **lineage**.

Tropisms	Growth movements in plants and nonmotile animals.
Vacuoles	Spherical droplets surrounded by a membrane, enclosing fat, protein, or other contents.
Vascular bundle	A grouping of **vascular tissues** in a plant.
Vascular plants	Plants with **vascular tissues (xylem** and **phloem).**
Vascular tissues	Plant tissues (**xylem** and **phloem**) that conduct liquids through the plant.
Vector	In genetics, a DNA sequence, such as a bacterial **plasmid**, that can easily replicate.
Vein	In animals, blood vessel carrying blood back toward the heart; in plants, extensions of **xylem** and **phloem** into the leaf.
Vertebrates	Animals with a backbone.
Virus	A fragment of nucleic acid, usually surrounded by protein, that can replicate only with the aid of intact cells.
Vitalist	One who views life as more than just a physical-chemical process.
Vitamin	An organic nutrient required in very small quantities.
Warm-blooded	Condition in which a high internal body temperature is maintained.
Xylem	Tissue in **vascular plants** that conducts water and dissolved minerals upward.
Zygote	A fertilized egg (always diploid).

INDEX

Please also consult the glossary on pages 128-150.